Visual Object Tracking using Deep Learning

This book covers the description of both conventional methods and advanced methods. In conventional methods, visual tracking techniques such as stochastic, deterministic, generative, and discriminative are discussed. The conventional techniques are further explored for multi-stage and collaborative frameworks. In advanced methods, various categories of deep learning-based trackers and correlation filter-based trackers are analyzed. The book also:

- Discusses potential performance metrics used for comparing the efficiency and effectiveness of various visual tracking methods.
- Elaborates on the salient features of deep learning trackers along with traditional trackers, wherein the handcrafted features are fused to reduce computational complexity.
- Illustrates various categories of correlation filter-based trackers suitable for superior and efficient performance under tedious tracking scenarios.
- Explores the future research directions for visual tracking by analyzing the real-time applications.

The book comprehensively discusses various deep learning-based tracking architectures along with conventional tracking methods. It covers in-depth analysis of various feature extraction techniques, evaluation metrics and benchmark available for performance evaluation of tracking frameworks. The text is primarily written for senior undergraduates, graduate students, and academic researchers in the fields of electrical engineering, electronics and communication engineering, computer engineering, and information technology.

Visual Object Tracking using Deep Learning

Ashish Kumar

CRC Press is an imprint of the
Taylor & Francis Group, an **informa** business

First edition published 2024
by CRC Press
2385 NW Executive Center Drive, Suite 320, Boca Raton, FL 33431

and by CRC Press
4 Park Square, Milton Park, Abingdon, Oxon, OX14 4RN

CRC Press is an imprint of Taylor & Francis Group, LLC

Library of Congress Cataloging-in-Publication Data
Names: Kumar, Ashish (Analyst), author.
Title: Visual object tracking using deep learning / by Ashish Kumar.
Description: First edition. | Boca Raton : CRC Press, 2024. | Includes bibliographical references and index.
Identifiers: LCCN 2023022866 (print) | LCCN 2023022867 (ebook) | ISBN 9781032490533 (hardback) | ISBN 9781032598079 (paperback) | ISBN 9781003456322 (ebook)
Subjects: LCSH: Automatic tracking. | Algorithms.
Classification: LCC TK6592.A9 K86 2024 (print) | LCC TK6592.A9 (ebook) | DDC 005.13--dc23/eng/20230520
LC record available at https://lccn.loc.gov/2023022866
LC ebook record available at https://lccn.loc.gov/2023022867

ISBN: 978-1-032-49053-3 (hbk)
ISBN: 978-1-032-59807-9 (pbk)
ISBN: 978-1-003-45632-2 (ebk)
ISBN: 978-1-032-59816-1 (ebk plus)

DOI: 10.1201/9781003456322

Typeset in Sabon
by SPi Technologies India Pvt Ltd (Straive)

Contents

Preface

Visual tracking is one of the fundamental problems of computer vision dealing with many real-time applications. It has a wide range of applications for improving the lifetime experience. Visual tracking can be used for tracking motions, pose, pedestrians, and autonomous vehicles, for improving living standards. Tracking can also be beneficial for crowd monitoring, security surveillance, and traffic monitoring for developing smart and safe cities for sustainable development.

Over the past few years, visual tracking algorithms are evolved from conventional frameworks to deep learning-based frameworks. Conventional trackers are fast in processing but are not effective for tedious environmental variations. Conventional trackers are explored as probabilistic, generative, discriminative, and deterministic frameworks. In a direction to improve tracking accuracy, the collaborative and multi-stage tracking frameworks are analyzed for real-time outcomes for tedious tracking problems.

Deep learning trackers provide discriminative features with efficient performance in complex tracking variations. It became necessary to extract the salient features in target appearance models using various deep neural networks. Handcrafted features such as color, texture, and HOG are integrated with deep features to improve their processing speed so that solutions to real-time tracking problems can be provided. Hyper-features are also extracted from various layers of deep neural network to address a wide range of tracking solutions.

Correlation filter trackers have gained momentum due to better accuracy and fast processing speed in comparison to conventional trackers and deep learning-based trackers. These trackers are suitable for pedestrian tracking, medical imaging, and traffic monitoring. Correlation filters-based trackers are suitable for handling dense occlusion as most of the trackers failed to recover when occlusion occurs in the scene.

Single object tracking is extended to multi-object tracking for exploring extended usage in tracking domain. Multi-object tracking is suitable for tracking autonomous vehicle tracking for driverless cars, traffic density monitoring, and pedestrian detection for preventing accidents and mishaps in the cities. Deep learning-based multi-object tracking provides many real-time applications appropriate for current scenarios. These classes of trackers make a remarkable contribution to the tracking field.

This book is an introductory attempt to provide in-depth analysis of recent advances in the field of visual tracking. It summarizes the salient features, advantages and limitations of various conventional as well as deep learning tracking frameworks that will be of interest to academicians, researchers, and scientists working in the domain. This book is a comprehensive and exhaustive analysis of the latest tracking algorithms and benchmark paramount for their practical deployment in real-time.

This book comprises twelve chapters, with exhaustive analysis of the latest visual tracking algorithms. Chapter 1, titled "Introduction to visual tracking in video sequences", explores the motivation and overview of the various applications of visual tracking in video sequences. Chapter 2, titled "Background and research orientation for visual tracking appearance model: Standards and models", classifies the tracking challenges along with various conventional and deep learning frameworks. Chapter 3, titled "Saliency feature extraction for visual tracking", highlights the importance of handcrafted features and deep features extraction steps with their efficiency and limitations in the presence of environmental variations. Chapter 4, titled "Performance metrics for visual tracking: A qualitative and quantitative analysis", investigates the performance evaluation metrics for analyzing tracker's robustness whether ground truth is available or not. Chapter 5, titled "Visual tracking data sets: Benchmark for evaluation", highlights the salient features, video sequences description for single object tracking, long-term tracking, and multi-modal tracking. Chapter 6, titled "Conventional framework for visual tracking: Challenges and solutions", discusses the three recent conventional tracking algorithms for addressing complex occlusion and background clutters tracking challenges.

Chapter 7, titled "Stochastic framework for visual tracking: Challenges and solutions", elaborates the probabilistic tracking framework along with experimental validation of a tracker based on particle filter framework. Chapter 8, titled "Multi-stage and collaborative tracking model", analyzes the tracking performing for a feature-fusion-based tracker and the benefits of collaborating the generative and discriminative approaches. Chapter 9, titled "Deep learning-based visual tracking model: A paradigm shift", explores the tracking framework based on fusion of deep learning features and handcrafted features from various deep neural networks such as CNN, RNN, and LSTM. Chapter 10, titled "Correlation filter-based visual tracking model: Emergence and upgradation", highlights the benefits and limitations of deep learning-based correlation filter trackers for long-term visual tracking. Chapter 11, titled "Future prospects of visual tracking: Application-specific analysis", elaborates the requirement of pruning techniques to reduce the complexity of deep neural network for faster tracking solutions and explainable AI to identify the architecture reasons for tracking failures. Chapter 12, titled "Deep learning-based multi-object tracking: Advancement for intelligent video analysis", explores the penetration of tracking frameworks for pedestrian tracking, pose estimation, and autonomous vehicle tracking.

The motivation behind this book is to highlight the impact of visual tracking algorithms to real-time scenarios and provide a reference study material for the beginners as well as the enthusiastic students.

Ashish Kumar

Author bio

Dr. Ashish Kumar, Ph.D., is working as an assistant professor with Bennett University, Greater Noida, U.P., India. He worked as an assistant professor with Bharati Vidyapeeth's College of Engineering, New Delhi. (Affiliated to GGSIPU, New Delhi, India) from August 2009 to July 2022. He has completed his PhD in Computer Science and Engineering from Delhi Technological University (formerly DCE), New Delhi, India in 2020. He has received Best Researcher award from the Delhi Technological University for his contribution in the computer vision domain. He completed his MTech with distinction in Computer Science and Engineering from GGS Indraprastha University, New Delhi. He has published more than 25 research papers in various reputed national and international journals and conferences. He has published more than 15 book chapters in various Scopus indexed books. He has authored/edited many books in the AI and computer vision domains. He is an active member in various international societies and clubs, as well as being a reviewer with many reputed journals and in technical program committee of various national and international conferences. Dr. Kumar has also served as a session chair in many international and national conferences. His current research interests include object tracking, image processing, artificial intelligence, and medical imaging analysis.

Chapter 1

Introduction to visual tracking in video sequences

1.1 OVERVIEW OF VISUAL TRACKING IN VIDEO SEQUENCES

Visual tracking is imperative in computer vision due to various real-time applications in video processing [1], surveillance [2], medical imaging [3], pedestrian tracking [4], activity recognition [5], robotics [6], augmented reality [7], and many more [8, 9]. However, tracking an object in a video sequence is tedious due to dynamic environmental conditions which include pose variations, illumination variations, full or partial occlusion, similar background, noise in the video (dust, rain, snow, haze, etc.), and fast and abrupt object motion.

Generally, visual tracking problem has three components for tracking an object in a video sequence [10]. First, there is a requirement for a robust appearance model for object representation and feature extraction. Second, an update model is required for periodic and constant updates of the tracking model along with the target state estimation in a video frame addressing the environmental variations. Third, a search mechanism is needed for the selection of the best probable target state for the final state estimation. In order to address tracking problems effectively, a robust appearance model is required [11]. An efficient appearance model should be able to capture the variations in target appearance with changing environmental conditions [12]. In this direction, various tracking algorithms have been proposed under various categories such as conventional tracking and DL-based tracking approaches. Broadly, conventional tracking methods are categorized as stochastic approach [13–16], deterministic approach [17–20], generative approach [12, 21–24], and discriminative approach [25–28]. Also, multi-stage tracking models [29, 30] and an amalgam of generative and discriminative models [31, 32] were proposed for addressing the tracking problems.

Figure 1.1 illustrates the various tracking methods under conventional methods. In another line of research, DL-based tracking appearance models were also proposed for addressing the tracking problems [33–36]. This category of appearance model utilized either deep features [33, 35] or hierarchical features [34, 36] extracted from different layers of DL models.

DOI: 10.1201/9781003456322-1

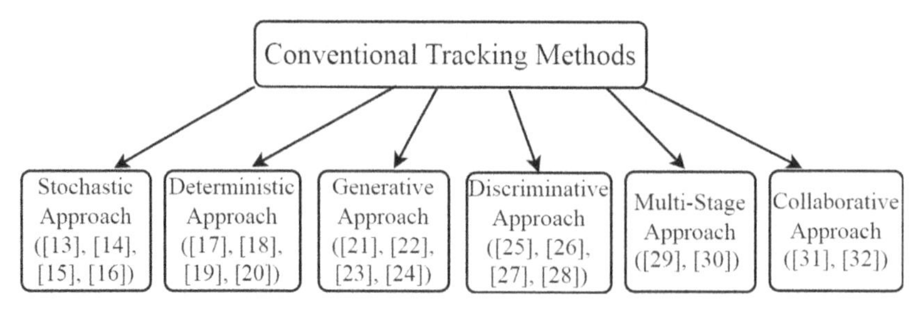

Figure 1.1 Categories of conventional tracking methods.

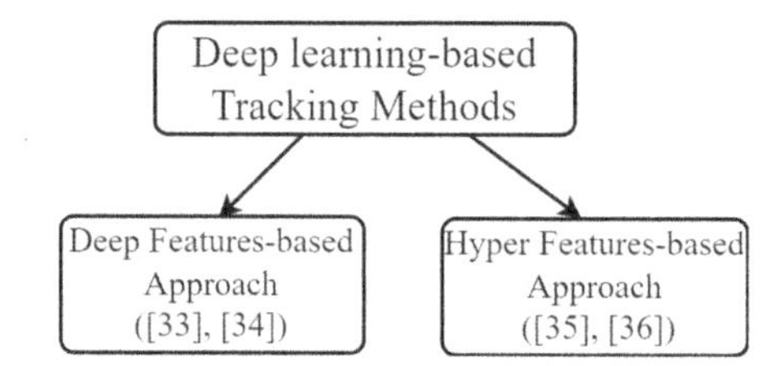

Figure 1.2 Categories of deep learning-based trackers.

Figure 1.2 illustrates the various DL-based approach. Apart from these categories, there is one more popular tracking algorithm category which is based on correlation filter (CF) [37, 38]. Under this category, context-aware CF trackers [39–41] and deep features-based CF trackers [9, 42, 43] are proposed to prevent tracking failures. This class of tracking algorithms provides significant good tracking results utilizing less computational power.

For a robust appearance model, tracking cues such as color, texture, edges, shape, gradient, audio, motion, infrared (IR), thermal, audio, deep features, etc. are extracted from the target. Earlier, a single feature was exploited from the target for building the tracker's measurement model [44, 45]. However, it has been emphasized that a single feature is not potential enough to cater the tracking failures during dynamic environmental variations. There is a requirement of exploiting multiple complementary cues in the tracker's appearance model to prevent the tracker's drift [46]. Complementary cues aim to address the limitations of each individual cue and enhance the tracker's performance during tough environmental conditions. These cues can be extracted from sensors such as vision, thermal, IR, and audio. Multiple cues can be extracted from a single sensor and fused into target appearance model [15, 29]. Also, multiple features from different sensors can be extracted to enhance the tracker's capabilities [17, 19]. Apart from these sensors, deep features can also be extracted from various DL models such as convolutional neural network (CNN) [36, 47], and recurrent neural network (RNN). Also, the hierarchical features extracted from multiple layers of a DL-model can be fused in a target appearance model [36, 47]. The next section will discuss the motivation and challenges in visual tracking.

1.2 MOTIVATION AND CHALLENGES

Visual tracking aims to keep track of the target's state in each frame of a video stream. For this, many robust appearance models have been proposed [9, 23, 37, 48]. However, visual tracking is still an open and challenging field that has attracted many researchers. Estimating the target's state in the presence of its deformation due to pose variations, illumination variations, fast motion, abrupt camera movement, and scale variations is tedious. Most of the proposed work has limited capability to address more than one environmental variation concurrently. Also, some of the proposed works have efficient performance but are computationally costly. Their real-time realization is still not possible due to either their slow speed or hardware constraints. There is a requirement for a robust tracker which not only has superior performance but is also applicable to real-time problems. For this, there is a robust tracking appearance model is needed, which can be motivated by the fusion of multiple cues which can address the numerous tracking challenges and enhance the tracking performance. It has been accepted that trackers with multiple features along with online cue reliability estimation have superior tracking results during tracking challenges.

In another line of research, tracking is also explored in unmanned vehicle (UAV) videos and videos captured by satellite [8, 9]. Tracking an object in a UAV video is challenging due to the fast motion, blur motion, occlusion, and out-of-view of the target. UAV devices, power consumption, and limited pay-load capability need to be checked for efficient tracking results. In addition, the video is captured from the top view, and the size of the target is relatively small. The actual size of the target is quite large in comparison to the one that is captured in the video. Similarly, tracking moving objects in a satellite video is quite tedious. Tracking challenges include the large field of vision, out-of-view, in-plane, and out-of-plane rotations of the target. The low contrast qualities of the UAV and satellite videos make it hard to discriminate the target from the background. Also, multiple objects captured in the scene, the small-scale size of the target, and target occlusion by the other objects, are the reasons that make tracking difficult in these videos in comparison to other ordinary videos. These challenges motivate researchers to keep exploring the tracking domain to provide robust solutions.

1.3 REAL-TIME APPLICATIONS OF VISUAL TRACKING

Visual tracking has numerous real-time applications in various fields. Some of the potential applications of visual tracking are discussed, as follows:

(a) Vehicle surveillance: The vehicle surveillance system [2] monitors the movement of the target on the road. In urban areas, on-road monitoring of the vehicle is important for the user's security and safety. Vehicle navigation is not only helpful in reducing road accidents, but also aids in minimizing the manual efforts of the monitoring team. Users can also be updated with the real-time parking situation for a preferred parking location. In addition, this technique can also be utilized for the automatic issuance of challan in case of traffic rule violations.

(b) Medical imaging: Visual tracking has also gained momentum in the medical field for robot-assisted laparoscopic surgeries [3]. During laparoscopic surgery, a camera is inserted in the patient's body, providing a real-time display of the internal organs on the monitor. In robot-assisted surgeries, visual tracking can aid in the automatic tracking of camera movement. This will improve the success of the procedure by automating the process of human-machine interaction.

(c) Pedestrian tracking: Aims to surveillance the target movement and ensure security [4]. This tracking is suitable for smart city projects which capture the real-time movement of the targets. It will provide the real-time trajectory of the target in a particular region. This will aid in automatic monitoring and hence has the potential to reduce crime rates.

(d) Activity recognition: This system aims to track the specific movement of the target from the data captured by the sensors [5]. The data will be helpful in tracking the target's fitness records and health monitoring. If there is any deviation in health data recorded, immediate health support can be extended to the target for timely treatment.

(e) Robotics: This domain deals with intelligent surveillance systems [6]. For real-time tracking and monitoring, data need to be captured from multiple sensors. However, it is not possible to analyze the captured data manually. For automatic and quick support, robotic-based tracking systems are proposed. These collaborative tracking systems have sufficient potential to capture human behavior and recognize the human activity.

(f) Augmented reality (AR): AR devices such as head-mounted display demands eye-tracking for gaze-based interaction [7]. For human visual attention tracking of eye movement is needed.

Apart from these domains, visual tracking can also be used for tracking moving objects in satellite video [8] and UAV tracking [9]. Also, it can be used for fault diagnosis in rotating machine gears [33]. In [8], small moving targets in satellite video were captured for applications in defense and military domains. The focus is to track the real-time movement of aircraft, rockets, missiles, and vehicles. This will be helpful to provide a more robust solution to the security of a nation. On the other hand, UAV tracking deals with long-term tracking of the object in the scene [9]. This will be helpful for crop monitoring, surveillance systems, remote sensing systems, intelligent transportation system, and video analysis. UAV tracking is widely used for automatic navigation along with driverless autonomous systems. This requires real-time monitoring and decision systems for successful tracking results.

1.4 EMERGENCE FROM THE CONVENTIONAL TO DEEP LEARNING APPROACHES

Over the last decade, visual tracking has attracted many researchers, academicians, scientists, and developers. The wide range of tracking applications and challenges has motivated people to provide progressive and robust solutions. Developing a robust appearance model is still an open and challenging problem of visual tracking.

Earlier tracking appearance models exploited conventional approaches for developing trackers. For the last few years, there has been a paradigm shift from conventional approaches to DL-based approaches. Conventional models exploited handcrafted features such as color, gradient, texture, IR, thermal, and audio in the trackers. However, DL-based trackers utilized either salient deep features such as deep RGB, deep texture, or hierarchical features computed from different layers of DL architecture. Popular traditional tracking methods are based on methodologies such as similarity template search, local feature extraction, background discrimination, and posterior probability estimation. These techniques utilize particle filter (PF) [15], temporal-spatial constraints [19], mean shift [28], attribute filter [18], and joint sparse representation [24] for developing a robust appearance model. Trackers based on the PF framework can address the tracking challenges efficiently [15]. They are simple and easy to develop, and hence require less computational power. But these trackers are not able to track small targets, fast motion, and motion blur efficiently. Also, the limitations of PF such as sample impoverishment and degeneracy need to be catered to for enhancing the tracking performance. Mean shift-based trackers utilize target global information for developing measurement models [28]. It computes the similarity between the target image and sample candidate images. Mean shift algorithms are lightweight and can be integrated with other techniques such as a radial bias for effective tracking results. However, these trackers are not able to cater to the target deformations due to heavy occlusion. Also, trackers based on mean shift are not adaptive to varying window sizes in search space. Sparse-based trackers are based on a generative approach that searches for similar target reasons using local sparse or joint sparse methods [24]. The intrinsic relationship or correlation between the local features and global features is jointly represented for a robust appearance model. These trackers are able to address the tracking challenges such as illumination variations during heavy occlusion and background noise. Due to high computational complexity, these trackers have limited usage in practical applications. Discriminative trackers discriminate the target from the background by utilizing both foreground and background information [25, 29]. In this, graph-based trackers exploited the target's geometric information for tracking during target deformation [25]. Also, a discriminative classifier is used to separate the foreground template from the background template for final state estimation [29]. These trackers are relatively stable and achieve satisfactory tracking results. But the tracker performance degrades when it is tedious to separate the target from the background, i.e. during a similar background and/or occlusion of the target by the same color object.

Further, multi-stage [29, 30] and collaborative [31, 32] tracking models have been proposed under traditional tracking solutions. Multi-stage trackers focus on coarse-to-fine localization of the target's state for accurate estimation. Initially, the target's location is roughly estimated during the first stage. The precise localization is done during the final stage of the estimation. Multi-stage trackers are efficient to address real-time tracking challenges. But fast-moving small targets and motion-blur impact the tracking performance. Similarly, for coarse-to-fine estimation, the limitations and benefits of both generative and discriminative trackers are integrated to propose hybrid trackers. Tracker's generative appearance model was constructed from the target's temporal and spatial features [31]. The discriminative information was fused using Fourier transform for expediting the tracking process. In [32], the generative appearance model was developed using the subspace learning method. The confidence score was adopted for positive and negative templates to discriminate the foreground information from the background. Hybrid trackers are efficient to address complex tracking scenarios but are not adaptive to the target's scale variations. In addition, the computational speed of these trackers is also extremely slow.

In another line of research, DL-based trackers have been proposed to address the limitations of conventional trackers [33–36]. DL-based trackers exhibited either deep features or hierarchical features to encode target information. Deep features are more robust in tracking variations in comparison to handcrafted features. In addition, deep features have the target's multi-level information essential for building a strong appearance model. In [33], CNN was used for automatic feature extraction. CNN was applied to thermal images for fault diagnosis in rotating machines. However, Li et al. [35] proposed multiple image patches corresponding to multiple complementary cues as the learning features for CNN. An online adaptive update strategy was also proposed to keep the tracker consistent with the changing scenarios. In [34], hierarchical feature maps were extracted from the hyper-Siamese network. The idea was to provide a trained end-to-end connected network for the extraction of comprehensive multiple features. In these ways low-level features were extracted from multiple layers for a better characterization of the target from the background. Zhang et al. [36] proposed a dynamic Siamese network for efficient tracking. Two fully connected Siamese networks were utilized for strong feature representation of RGB and thermal images independently. The tracker has shown good tracking performance but requires a lot of training data.

Recently, CF-based trackers [26, 37, 42, 47] have achieved great popularity in object tracking. These trackers have shown impressive performance in the presence of tedious environmental variations. Under this, authors have proposed either context-aware or deep features-based CF tracking approaches. Context-aware-based CF trackers search for the target position in an image patch consisting of both information, i.e. target as well as background [39]. To suppress the background information an efficient classifier was used along with circular shifting. Holistic and local part information about the target was adaptively weighted and integrated to obtain the target's state in the scene. Similarly, the authors proposed a patch-based method for the tracking process [40]. Patches were distributed over the whole image to differentiate the target from the background. The distribution of reliable patches was determined by utilizing the reliability metric along with the probability model. A voting scheme was employed for the final estimation of the target's state. Context-aware CF trackers have shown effective and efficient performance during background noise and heavy occlusion. However, the performance of these trackers degrades during large-scale variations and incorrect target selection in complex videos. To address this, kernelized CF-based trackers are proposed to cope with significant deformation in target appearance due to these challenges. An adaptive scale calculation scheme was adopted to handle scale variations [43]. The conventional classifier output was used to obtain the accurate location of the target. However, the authors proposed Gaussian distribution response to obtain the correlation between the target image in the search area during the tracker's drift [42]. The scale adaptation was used to identify the accurate candidate target. In another direction, a correlation filter was used with thermal images to boost the tracker's performance during the night [47]. Multiple weak CF-based trackers corresponding to each CNN layer response were used for a strong representation of the target's spatial information. Finally, an ensemble technique was adopted to combine the response to obtain a strong response map for the target's state estimation. Scale adaptation technique was used to prevent the tracker's drift during scale variations. In summary, CF-based trackers have shown efficient and effective performance in the presence of dynamic environmental variations in complex videos.

1.5 PERFORMANCE EVALUATION CRITERIA

To evaluate the practical applicability of the tracker, it is necessary to test the robustness and reliability of the tracker on standardized performance metrics and publicly available good quality benchmark [11]. Different authors have utilized different approaches to provide accurate and reliable trackers. A lot of work has been published in single object tracking from conventional model to DL-based models. However, the performance evaluation criteria are heterogeneous in the papers. There is no common way to assess the superiority of the published work in order of their capability to address the real-time tracking variations.

The existing performance criteria are arbitrary in nature. Authors compare their work against a limited set of state-of-the-art exploiting limited number of performance metrics [45]. There exists an abundance of standardized performance metrics in various papers. However, there is no homogeneity in the experimental validation of the tracker by various researchers. The reason may be that the authors either are not aware of those standardized evaluation criteria, or that they chose certain metrics in which the tracker performance is effective and efficient. To bridge this gap, we have investigated recent, effective, and popular performance metrics in this book. The elaborated performance metrics are suitable to analyze the systematic experimental evaluation of the tracking algorithms. These metrics do not specify the superiority or ranking of the existing work but rather help in understanding the robustness of the tracker's algorithmic design and interpretability of the tracking results [49].

In a similar line, there is a requirement of a good quality data set to check the functionality of the tracker in synthetic environment before deployment in the real world. For this, the trackers must be evaluated on public data sets, which are rich in tedious tracking challenges. The data set must contain suitable length of videos so that rigorous qualitative analysis of the tracker can be performed. Also, the data set must have sufficient number of videos with varying lighting and scene variations so that actual precision and accuracy of the tracker's response can be experienced. It is advisable to evaluate the tracking algorithms on multiple data sets so that biasness and overfitting can be avoided [29]. Public data sets should be preferred over self-generated data to compare the tracker's performance in similar scenarios.

1.6 SUMMARY

In this chapter, we have summarized the overview of the environmental challenges and the significant problems of object tracking. Object tracking methods are categorized into conventional methods and DL-based methods. The limitations and benefits of tracking approaches under each category are discussed in detail. It has been inferred that conventional trackers have shown superior performance during complex environmental conditions. But their real-time implementation is partially explored. On the other hand, DL-based trackers are better in terms of accuracy and performance for practical application. However, their requirements for the large training data set and specialized hardware restrict their usability for real-world problems. Further, effective and efficient tracking methods under CF are highlighted in terms of their achievement in object tracking. These trackers can address the tracker's drift during object complex deformation. The potential real-time applications of visual tracking are elaborated to generate the researchers' interest in the domain. The practical applications of tracking methods from medical imaging to fast-moving UAV tracking are discussed to highlight the challenges.

REFERENCES

1. Song, D., C. Kim, and S.-K. Park, A multi-temporal framework for high-level activity analysis: Violent event detection in visual surveillance. *Information Sciences*, 2018. **447**: pp. 83–103.
2. Messoussi, O., F.G.D. Magalhães, F. Lamarre, F. Perreault, I. Sogoba, G.-A. Bilodeau, and G. Nicolescu. Vehicle detection and tracking from surveillance cameras in urban scenes. in *International Symposium on Visual Computing*. 2021. Springer.
3. Zhang, X. and S. Payandeh, Application of visual tracking for robot-assisted laparoscopic surgery. *Journal of Robotic systems*, 2002. **19**(7): pp. 315–328.
4. Wang, N., Q. Zou, Q. Ma, Y. Huang, and D. Luan, A light tracker for online multiple pedestrian tracking. *Journal of Real-Time Image Processing*, 2021. **18**(1): pp. 175–191.
5. Krahnstoever, N., J. Rittscher, P. Tu, K. Chean, and T. Tomlinson. Activity recognition using visual tracking and RFID. in *2005 Seventh IEEE Workshops on Applications of Computer Vision (WACV/MOTION'05)-Volume 1*. 2005. IEEE.
6. Ahmed, I., S. Din, G. Jeon, F. Piccialli, and G. Fortino, Towards collaborative robotics in top view surveillance: A framework for multiple object tracking by detection using deep learning. *IEEE/CAA Journal of Automatica Sinica*, 2021. **8**(7): pp. 1253–1270.
7. Kapp, S., M. Barz, S. Mukhametov, D. Sonntag, and J. Kuhn, ARETT: Augmented reality eye-tracking toolkit for head-mounted displays. *Sensors*, 2021. **21**(6): p. 2234.
8. Pei, W. and X. Lu, Moving object tracking in satellite videos by Kernelized correlation filter based on color-name features and Kalman prediction. *Wireless Communications and Mobile Computing*, 2022. **2022**: pp. 1–16.
9. Kumar, A., R. Jain, V. A. Devi, & A. Nayyar, (Eds.). *Object Tracking Technology: Trends, Challenges, Impact, and Applications*, 2023. Springer.
10. Liu, Y., F. Yang, C. Zhong, Y. Tao, B. Dai, and M. Yin, Visual tracking via salient feature extraction and sparse collaborative model. *AEU-International Journal of Electronics and Communications*, 2018. **87**: pp. 134–143.
11. Granstrom, K., M. Baum, and S. Reuter, Extended object tracking: Introduction, overview and applications. arXiv preprint arXiv:1604.00970, 2016.
12. Dou, J., Q. Qin, and Z. Tu, Robust visual tracking based on generative and discriminative model collaboration. *Multimedia Tools and Applications*, 2017. **76**(14): pp. 15839–15866.
13. Cai-Xia, M. and Z. Xin-Yan, Object tracking method based on particle filter of adaptive patches combined with multi-features fusion. *Multimedia Tools and Applications*, 2019. **78**(7): pp. 8799–8811.
14. Firouznia, M., K. Faez, H. Amindavar, and J.A. Koupaei, Chaotic particle filter for visual object tracking. *Journal of Visual Communication and Image Representation*, 2018. **53**: pp. 1–12.
15. Kumar, A., G.S. Walia, and K. Sharma, Real-time visual tracking via multi-cue based adaptive particle filter framework. *Multimedia Tools and Applications*, 2020. **79**(29): pp. 20639–20663.
16. Walia, G.S., A. Kumar, A. Saxena, K. Sharma, and K. Singh, Robust object tracking with crow search optimized multi-cue particle filter. *Pattern Analysis and Applications*, 2020. **23**(3): pp. 1439–1455.

17. Ma, J., P. Liang, W. Yu, C. Chen, X. Guo, J. Wu, and J. Jiang, Infrared and visible image fusion via detail preserving adversarial learning. *Information Fusion*, 2020. **54**: pp. 85–98.
18. Mo, Y., X. Kang, P. Duan, B. Sun, and S. Li, Attribute filter based infrared and visible image fusion. *Information Fusion*, 2021. **75**: pp. 41–54.
19. Xiao, J., R. Stolkin, Y. Gao, and A. Leonardis, Robust fusion of color and depth data for RGB-D target tracking using adaptive range-invariant depth models and spatio-temporal consistency constraints. *IEEE Transactions on Cybernetics*, 2017. **48**(8): pp. 2485–2499.
20. Zhang, J., X. Yang, Y. Fu, X. Wei, B. Yin, and B. Dong. Object tracking by jointly exploiting frame and event domain. in *Proceedings of the IEEE/CVF International Conference on Computer Vision*. 2021.
21. Gao, X., Y. Zhou, S. Huo, Z. Li, and K. Li, Robust object tracking via deformation samples generator. *Journal of Visual Communication and Image Representation*, 2022. **83**: p. 103446.
22. Liu, G., Robust visual tracking via smooth manifold kernel sparse learning. *IEEE Transactions on Multimedia*, 2018. **20**(11): pp. 2949–2963.
23. Wang, T., Z. Ji, J. Yang, Q. Sun, and P. Fu, Global manifold learning for interactive image segmentation. *IEEE Transactions on Multimedia*, 2020. **23**: pp. 3239–3249.
24. Wang, Y., X. Luo, L. Ding, and S. Hu, Visual tracking via robust multi-task multi-feature joint sparse representation. *Multimedia Tools and Applications*, 2018. **77**(23): pp. 31447–31467.
25. Du, D., H. Qi, L. Wen, Q. Tian, Q. Huang, and S. Lyu, Geometric hypergraph learning for visual tracking. *IEEE Transactions on Cybernetics*, 2017. **47**(12): pp. 4182–4195.
26. Hu, Q., Y. Guo, Z. Lin, W. An, and H. Cheng, Object tracking using multiple features and adaptive model updating. *IEEE Transactions on Instrumentation and Measurement*, 2017. **66**(11): pp. 2882–2897.
27. Phadke, G. and R. Velmurugan, Mean LBP and modified fuzzy C-means weighted hybrid feature for illumination invariant mean shift tracking. *Signal, Image and Video Processing*, 2017. **11**(4): pp. 665–672.
28. Rowghanian, V. and K. Ansari-Asl, Object tracking by mean shift and radial basis function neural networks. *Journal of Real-Time Image Processing*, 2018. **15**(4): pp. 799–816.
29. Kumar, A., G.S. Walia, and K. Sharma, A novel approach for multi-cue feature fusion for robust object tracking. *Applied Intelligence*, 2020. **50**(10): pp. 3201–3218.
30. Walia, G.S., S. Raza, A. Gupta, R. Asthana, and K. Singh, A novel approach of multi-stage tracking for precise localization of target in video sequences. *Expert Systems with Applications*, 2017. **78**: pp. 208–224.
31. Dai, M., S. Cheng, and X. He, Hybrid generative–discriminative hash tracking with spatio-temporal contextual cues. *Neural Computing and Applications*, 2018. **29**(2): pp. 389–399.
32. Wang, Y., X. Luo, L. Ding, and S. Hu, Multi-task based object tracking via a collaborative model. *Journal of Visual Communication and Image Representation*, 2018. **55**: pp. 698–710.
33. Choudhary, A., T. Mian, and S. Fatima, Convolutional neural network-based bearing fault diagnosis of rotating machine using thermal images. *Measurement*, 2021. **176**: p. 109196.

34. Kuai, Y., G. Wen, and D. Li, Hyper-Siamese network for robust visual tracking. *Signal, Image and Video Processing*, 2019. **13**(1): pp. 35–42.
35. Li, H., Y. Li, and F. Porikli, Deeptrack: Learning discriminative feature representations online for robust visual tracking. *IEEE Transactions on Image Processing*, 2015. **25**(4): pp. 1834–1848.
36. Zhang, X., P. Ye, S. Peng, J. Liu, and G. Xiao, DSiamMFT: An RGB-T fusion tracking method via dynamic Siamese networks using multi-layer feature fusion. *Signal Processing: Image Communication*, 2020. **84**: p. 115756.
37. Yuan, D., W. Kang, and Z. He, Robust visual tracking with correlation filters and metric learning. *Knowledge-Based Systems*, 2020. **195**: p. 105697.
38. Yuan, D., X. Zhang, J. Liu, and D. Li, A multiple feature fused model for visual object tracking via correlation filters. *Multimedia Tools and Applications*, 2019. **78**(19): pp. 27271–27290.
39. Chen, K., W. Tao, and S. Han, Visual object tracking via enhanced structural correlation filter. *Information Sciences*, 2017. **394**: pp. 232–245.
40. Li, Y., J. Zhu, and S.C. Hoi. Reliable patch trackers: Robust visual tracking by exploiting reliable patches. in *Proceedings of the IEEE Conference on Computer Vision and Pattern Recognition*. 2015.
41. Liu, T., G. Wang, and Q. Yang. Real-time part-based visual tracking via adaptive correlation filters. in *Proceedings of the IEEE Conference on Computer Vision and Pattern Recognition*. 2015.
42. Li, C., X. Liu, X. Su, and B. Zhang, Robust kernelized correlation filter with scale adaption for real-time single object tracking. *Journal of Real-Time Image Processing*, 2018. **15**(3): pp. 583–596.
43. Lian, G.-Y., A novel real-time object tracking based on kernelized correlation filter with self-adaptive scale computation in combination with color attribution. *Journal of Ambient Intelligence and Humanized Computing*, 2020: pp. 1–9.
44. Narayana, M., H. Nenavath, S. Chavan, and L.K. Rao, Intelligent visual object tracking with particle filter based on Modified Grey Wolf Optimizer. *Optik*, 2019. **193**: p. 162913.
45. Walia, G.S. and R. Kapoor, Intelligent video target tracking using an evolutionary particle filter based upon improved cuckoo search. *Expert Systems with Applications*, 2014. **41**(14): pp. 6315–6326.
46. Kumar, A., G.S. Walia, and K. Sharma, Recent trends in multi-cue based visual tracking: A review. *Expert Systems with Applications*, 2020. **162**: p. 113711.
47. Liu, Q., X. Lu, Z. He, C. Zhang, and W.-S. Chen, Deep convolutional neural networks for thermal infrared object tracking. *Knowledge-Based Systems*, 2017. **134**: pp. 189–198.
48. Yuan, D., X. Li, Z. He, Q. Liu, and S. Lu, Visual object tracking with adaptive structural convolutional network. *Knowledge-Based Systems*, 2020. **194**: p. 105554.
49. Čehovin, L., A. Leonardis, and M. Kristan, Visual object tracking performance measures revisited. *IEEE Transactions on Image Processing*, 2016. **25**(3): pp. 1261–1274.

Chapter 2

Research orientation for visual tracking models

Standards and models

2.1 BACKGROUND AND PRELIMINARIES

Over the last decade, visual tracking has gained momentum in research. Wide progress in science and technology and the real-time applications of visual tracking have attracted many researchers. Visual tracking can be used to track the moving target in a video in the presence of tracking challenges. Tracking challenges can be broadly classified into system-related challenges and environmental-related challenges.

Figure 2.1 illustrates the classification of tracking challenges.

System-related challenges deal with the hardware limitation of the sensors and cameras used for capturing the video. The lens quality, camera/sensor calibration (CSC), resolution (CSR), and camera location (CLC) are the main challenges to be addressed for obtaining a quality video suitable for visual tracking. On the other hand, environmental challenges include variations due to target movement from one frame to another. Various visual tracking challenging attributes namely, illumination variation (IV), scale variations (SV), deformation (DEF), occlusion (OCC), background clutters (BC), out-of-view (OV), object rotation (OR), low resolution (LR), fast motion (FM), and motion blur (MB) are listed in [1]. During UAV tracking,

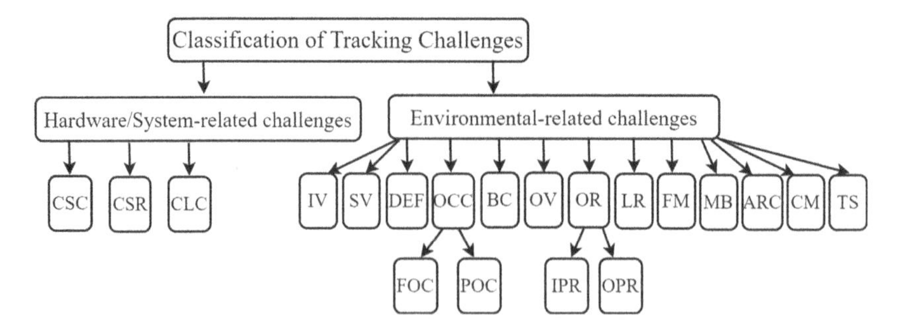

Figure 2.1 Classification of various tracking challenges.

DOI: 10.1201/9781003456322-2

the challenging attributes such as aspect ratio change (ARC), fast camera motion (CM), and small target size (TS) also need to be addressed [2].

Figure 2.2 represents the various tracking challenges in the video frames taken from video benchmark data sets [1, 2]. IV challenge refers to the abrupt change in illumination in the target region. This change will impact the intensity, brightness, and contrast of the target in the scene. Hence, the color feature extracted from the vision sensor is inefficient in developing a robust appearance model. When the target's bounding box in the current video frame varies above a threshold from the initial frame, it leads to SV challenge. These variations lead to a change in the number of pixels in the target region. However, DEF determines the non-rigid change in the target's shape in the current frame than the initial frame. OCC can be either partial occluded (POC) or fully occluded (FOC). When these challenges occur in the scene, trackers need to re-detect the target in the future frame for continuous tracking. When the target background is of similar color to that of the target, then it leads to BC. It is tedious to discriminate the target from the background when both of them are alike in color. OV challenge occurs when the target is either at the frame boundary or some portion of the target is clipped off from the scene. Similar to OCC, OV also requires re-detection after this challenge. OR refers to rotational challenges either in-plane (IPR) or out-of-plane rotation (OPR). IPR and OPR occur in the frame when the target moves in or out of the image plane, respectively. LR occurs in the image frame when the number of pixels is below a particular threshold. When the target is moving fast and changing its location frequently, it is referred to as FM. However, when both the target and camera are moving the lead to MB, the target in the frame became blurred due to the shaky camera. ARC refers to when the bounding box aspect ratio is below a specific threshold. Finally, CM and VPC refer to abrupt camera movement and change in the target's viewpoint. These tracking challenges should be addressed for developing an effective appearance model. The next section will detail the various appearance model under conventional tracking frameworks.

2.2 CONVENTIONAL TRACKING METHODS

The appearance model under conventional tracking methods is based on handcrafted features. Handcrafted features are extracted from the vision camera, IR, thermal, and audio sensors. The framework based on probabilistic, mean shift (MS), fragment-based, image patches approaches are categorized under conventional approaches. Generally, these frameworks are termed stochastic, deterministic, generative, and discriminative based on the methodology exploited in the appearance model. The salient features of these frameworks are shown in Table 2.1. The details under each category are discussed as follows.

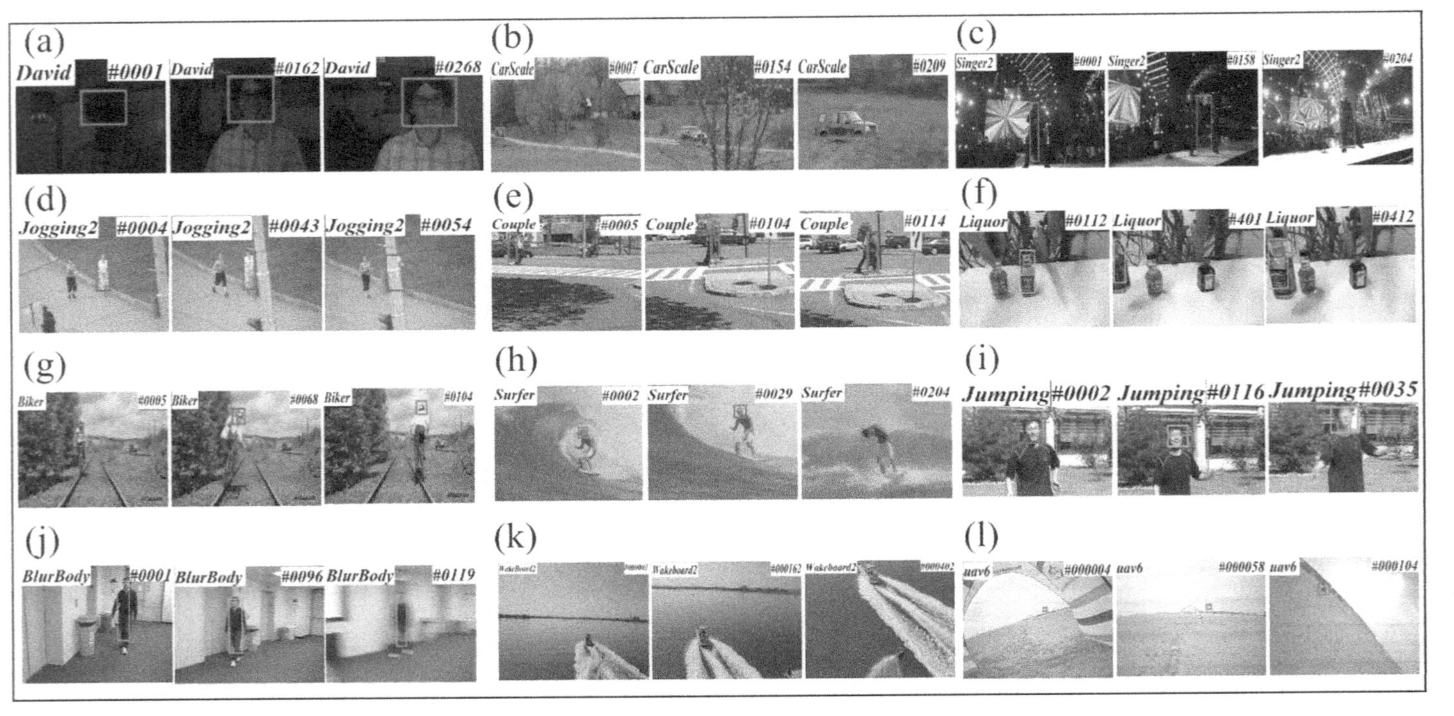

Figure 2.2 Sample video frames from benchmark data sets [1, 2], representing the various environmental challenges (a) Illumination variations (IV), (b) Scale variations (SV), (c) Deformation (DEF), (d) Occlusion (OCC), (e) Background clutters (BC), (f) Out-of-view (OV), (g) Object rotation (OR), (h) Low resolution (LR), (i) Fast motion (FM), (j) Motion blur (MB), (k) Aspect ratio change (ARC), and (l) Fast camera motion (CM) and Small target size (TS). Each sample video frame is labeled with sequence name and frame number on the left and right corners, respectively.

Table 2.1 Salient feature of various conventional tracking frameworks

SN	*Attributes*	*Stochastic approach [3–6]*	*Deterministic approach [7–10]*	*Generative approach [11–14]*	*Discriminative approach [15–18]*	*Multi-stage approach [19–22]*	*Hybrid approach [23–26]*
1	Methodology	Probabilistic methods, PF-based methods, Kalman filter-based methods	Mean shift, fragment-based methods	Sparse learning, template matching	Graph-based methods, discriminative classifier	Coarse-to-fine estimation	Combination of generative and discriminative approach
2	Handcrafted features	✓	✓	✓	✓	✓	✓
3	Multi-cue features	✓	✓	✓	✓	✓	✓
4	Feature fusion	✓	✓	✓	✓	✓	✓
5	Multiple data sets	✓	✓	✓	✓	✓	✓

2.2.1 Stochastic approach

Under this approach, probabilistic tracking methods based on Monte Carlo simulation and Bayesian framework have been explored. PF, Kalman filter, and extended Kalman filter were exploited for developing robust tracking solutions. Among these, PF has been widely explored due to its ease of implementation and potential to handle nonlinear and non-Gaussian tracking problems efficiently. Multiple features are extracted from the target to maintain tracker efficiency during environmental variations. In this direction, Kumar et al. [3] have exploited color, linear binary pattern (LBP), and pyramid of the histogram of gradient (PHOG) in the tracker's appearance model in the PF framework. Rank-based adaptive score fusion is used for feature integration. Outlier detection mechanism detects the unimportant particles to reduce the tracker's computations. In [4], the authors have utilized color features in the measurement model. Experimental results claim that the tracker converges faster to better results in fewer iterations. Taylor et al. [5] have utilized the Kalman filter to improve tracking efficiency by cropping the image during detection. Authors have exploited fuzzy-based fusion of features along with Markov process-based outlier detection to improve tracker processing during tough environmental challenges [6]. However, Cai-Xai et al. [27] have utilized fuzzy C-Means to reduce the number of processing particles. The tracker can address the impact of similar targets and occlusion effectively. Multi-modal information from sensors such as audio, depth, and thermal are integrated with vision sensors to provide effective tracking solutions [28, 29]. Table 2.2 shows the representative work proposed under stochastic tracking frameworks.

2.2.2 Deterministic approach

The deterministic approach exploited the MS framework search for local maxima in the space for efficient real-time tracking. For this, the authors have exploited three sub-appearance models for each considered feature and fused them in an MS framework [7]. Local estimation is done using MS, and global estimation was conducted to reduce the number of iterations in the tracker for the final estimation of the target. Phadke et al. [8] have modified the color and texture features for developing an appearance model invariant to IV and scale changes. A hybrid feature is generated using MS for better localization and robust tracking. In [9], authors have proposed to cluster two independent trackers using fuzzy C-means clustering. The combined tracker is efficient to address sudden deformations, movement, and occlusion. Yu et al. [10] have proposed 3-D spatial histograms to adapt to scale variations. Multi-scale has suppressed the background information and discriminated against the target effectively. In [30], authors have integrated color and modified texture features (uniform interlaced derivative pattern) to address the limitations of MS trackers. Also, scale adaption strategy is

Table 2.2 Representative work under stochastic approach

SN	*Reference*	*Sensor*	*Framework*	*Extracted feature*	*Summary*
1	Kumar et al. [3]	Vision	PF	Color, LBP, PHOG	Adaptive rank-based fusion and outlier detection mechanism to improve tracker's efficiency.
2	Narayana et al. [4]	Vision	PF	Color	Particles are optimized before the resampling techniques for precise state estimation.
3	Taylor et al. [5]	Vision	Kalman filter	Color	Kalman filter is used to crop the image for faster processing.
4	Walia et al. [6]	Vision	PF	Color, LBP	Fuzzy-based adaptive fusion of features to create clear decision boundaries for selecting unimportant particles.
5	Cai-Xai et al. [27]	Vision	PF	Color, HOG	Adaptive fusion of each sub-patch to address target deformation and divergence problems.
6	Walia et al. [28]	Vision, Thermal	PF	Color, Thermal, Motion	Adaptive fusion of features at particle level to improve tracking quality.
7	Liu et al. [29]	Audio, depth	PF	Audio, Depth	Combined audio and depth features to address the multiple occlusions in the scene.

proposed to change the scale according to the target's size. The authors have integrated the HSV color and texture feature along with four neighborhood strategy to address the partial occlusion and similar background problems [31]. However, the authors have integrated depth information along with color to efficiently handle the occlusion [32]. Temporal consistency and spatial consistency are considered to adaptively estimate the target location. Further, the salient details of the representative work using the MS framework are shown in Table 2.3. The next section will detail the generative tracking approach.

Table 2.3 Representative work under deterministic approach

SN	*Reference*	*Sensor*	*Framework*	*Extracted feature*	*Summary*
1.	Dhassi et al. [7]	Vision	MS	Color Correlogram, Edge Orientation Histogram, Uniform LBP	Global state and local state were computed for each state for final state estimation.
2.	Phadke et al. [8]	Vision	MS	Mean LBP, modified fuzzy C-Means weighted color histogram	Hybrid and modified color and texture features are proposed to address IV and SV.
3.	Rowghanian et al. [9]	Vision	MS	Color, Spatial, Corrected background weighted histogram	Two independent trackers i.e. MS and radial bias function integrated to produce a confidence map.
4.	Yu et al. [10]	Vision	MS	Spatial histogram, multi-scale model	Weighted background strategy is used to separate the target from the background efficiently.
5.	Razavi et al. [30]	Vision	MS	Color, Texture	Adaptive resize window was considered to adapt to varying target scales and sizes.

(*Continued*)

Table 2.3 (Continued)

SN	Reference	Sensor	Framework	Extracted feature	Summary
6.	Liu et al. [31]	Vision	MS	HSV color, Texture	Four neighborhood search method was proposed for enhancing tracking results.
7.	Xiao et al. [32]	Vision, Depth	MS	RGB, Depth	Spatio-temporal consistency for both local and global models for handling occlusion.

2.2.3 Generative approach

Generative-based trackers search for the most similar region to the target region for final state estimation. The salient features of the work proposed under this category are tabulated in Table 2.4. In this direction, authors have utilized patch-based sparse coding coefficient histograms to determine similarity and address the occlusion during tracking [11]. Quality assessment algorithms and context information are also included during similarity computation to enhance the tracking results. However, Hu et al. [12] have fused the two-channel information obtained from the binocular camera along with the geometry constraints for long-term tracking. A stereo filter is used to check the consistency between the two channels to handle false detections during tracking. In [14], authors have proposed a weighted temporal correlation between the consecutive frames to maintain the tracker's stability. However, the tracker fails to handle complex object translation and scale changes effectively. To address this, multi-view group similarity projections between reliable and unreliable features are presented in [13]. A nonlocal regularizer is introduced for group projections to enhance the tracking performance. Similarly, Wang et al. [33] have represented multi-task sparse learning for each feature individually. Outliers are detected to improve the tracker's processing during encountered challenges. In another direction, multi-modal visual information is integrated to propose a robust tracker under extremely complex environmental situations [34, 35]. In [34], two image patches for each feature are concatenated into a linear feature vector for object representation. The image thermal modularity integration with grayscale information has improved the tracker's performance during nonzero illumination variations and similar background situations. However, the integration of depth with the RGB color feature has handled the tracker's efficiency during heavy occlusion situations [35]. Depth information in the appearance model ensures the timely update of the augmented dictionary

Table 2.4 Representative work under the generative approach

SN	*Reference*	*Sensor*	*Framework*	*Extracted feature*	*Summary*
1.	Feng et al. [11]	Vision	Patch-based sparse representation	Temporal and Spatial context	Weighted similarity between the candidate patch and the target patch determined in the target's appearance model.
2.	Hu et al. [12]	Binocular vision	Stereo filter	Motion	Combined the target's information obtained from two channels by the binocular camera.
3.	Liu [14]	Vision	Kernel sparse learning	Covariance matrix descriptor	Proposed online adaptive dictionary update to facilitate the target sudden deformations.
4.	Kang et al. [13]	Vision	Sparse representation	Color, Shape, and Texture	Multi-view information corresponding to each cue extracted fused using an online discriminant matrix.
5.	Wang et al. [33]	Vision	Joint sparse representation	Color, Texture, and edge	Multi-task features were sparsely represented for the adaptive feature template.
6.	Li et al. [34]	Vision, and Thermal	Laplacian sparse representation	Grayscale, and Thermal	Deployed similarity between the two modalities to refine target representation.
7.	Ma and Xiang [35]	Vision, and Depth	Sparse learning	RGB and Depth	Depth information is incorporated into the template for model update during occlusion.

during heavy occlusion to maintain the tracker's stability. The next category of traditional tracking approaches i.e. discriminative approach is discussed in the next section.

2.2.4 Discriminative approach

The discriminative approach considers tracking as a binary classification problem to separate the target from the background. Discriminative trackers are efficient to address heavy occlusion and similar background challenges efficiently. In this direction, a geometric hypergraph is used to determine the strong geometric relationship in the parts of the target between different frames [15]. Confidence-aware sampling method is used to reduce noise in the graph to make the tracker computationally efficient. But the tracker's performance diminishes during fast motion and motion blur. To propose an efficient tracker, the authors have utilized automatic updates of the model based on the penalty factor [16]. This ensured the tracker is to be stable and flexible during tough environmental variations. Three distinct hypergraphs are constructed corresponding to each feature and integrated linearly with adaptive template constraints [17]. To enhance the tracker's discriminative capability, a dictionary of positive and negative templates is updated in accordance to target appearance variations. But the authors have used an iterative optimization technique to find optimal dictionary updates for better performance [18]. Nonlinear theory in a supervised environment to enhance the discriminative power of the tracker. In [36], the nonlinear relationship between the object parts is represented using the dense and hybrid structures of hypergraphs. Dense structures are optimized using an approximation algorithm for tracking the target. Authors have proposed iterative graph theory for selecting, matching, and estimating the target's part for tracking objects [37]. Structural information in the selection parts has improved the selection for efficient tracking. However, multi-modal information to address complex background problems is integrated in [38]. Textureless depth templates are updated using an adaptive update mechanism, which restricts the erroneous tracker's update and prevents drift during the target's appearance variations. Table 2.5 shows the salient feature of the trackers categorized under the discriminative approach. The details related to the multi-stage approach are as follows.

2.2.5 Multi-stage approach

Generally, the trackers under this category adopt a two-stage strategy to provide robust tracking algorithms. The procedure is to provide a coarse-to-fine estimation of the target state by ensuring the low computational complexity of the tracker.

In this direction, Kumar et al. [19] have proposed a multi-stage tracker from rough to precise estimation of the target's state. Two complementary

Table 2.5 Representative work under discriminative approach

SN	*Reference*	*Sensor*	*Framework*	*Extracted feature*	*Summary*
1.	Du et al. [15]	Vision	Geometric hypergraph learning	The geometric relationship between different parts of the image	Confidence-aware sampling method to cater to the tracking challenges.
2.	Hu et al. [16]	Vision	Linear interpolation	CN, HOG, and Adjusted local histogram	Update strategy based on penalty method to maintain tracker's stability.
3.	Lu et al. [17]	Vision	Probabilistic hypergraph learning	Spatial relationship, Local clustering information, and neighboring information in feature space	Dynamic update of positive and negative samples for the consistent update of the tracker during environmental variations.
4.	Ma et al. [18]	Vision	Nonlinear learning	CIE lab color feature and HOG	Iterative optimization technique used to determine the optimal updated adaptive dictionary.
5.	Li et al. [36]	Vision	Hybrid structure hypergraph	HSV color and LBP	Non-uniform hypergraph is used to model the interdependencies among various object targets.
6.	Du et al. [37]	Vision	Unified energy minimization	CN and HOG	Minimize energy function to determine the most efficient geometric graph.
7.	Chen et al. [38]	Vision, and Depth	Multiple instance learning	Texture and Depth	Efficient update scheme to eliminate incorrect model updates to prevent tracker's drift.

features, namely LBP and HOG, are fused using an adaptive fusion in the tracker's appearance model. To enhance the tracker's discriminative power, the sample dictionary is updated periodically and consistently. This update prevents the tracker's drift during appearance variations. However, the authors have proposed a coarse-to-fine tracker in which the target is roughly localized using optical flow in the initial stage [20]. For fine state estimation, three features such as RGB, texture, and PHOG are extracted to develop a robust appearance model. Context-sensitive cue reliability is extracted to adapt the tracker to environmental variations. In [21], authors leverage motion features and visual features for the target's precise estimation. Visual features, namely color and HOG, are fused adaptively using a fuzzy-based fusion model. The random forest-based classifier is exploited to determine the unreliable samples to prevent tracking failures. However, Wu et al. [22] have proposed a multi-stage tracker by integrating principal component analysis (PCA) and background alignment. Initially, the unreliable templates are removed to reduce the computational load. After this, the final state is estimated by integrating the basic vectors and the background template. The tracker is robust to occlusion and similar backgrounds. In [39], the authors roughly estimate the target using a motion feature and three complementary features for the final state estimation. Unified feature vector is generated by exploiting cross-diffusion in the adaptive update model. Transductive cue reliability ensures the tracker resistant to environmental variations. In summary, multi-stage trackers are efficient and effective to address dynamic tracking variations. These trackers utilize less computational load for real-time application of the algorithms.

2.2.6 Collaborative approach

In this section, visual tracking algorithms based on a collaborative approach are discussed. The trackers under this category exploit a collaborative approach for developing a robust appearance model. The strengths and benefits of both approaches such as generative and discriminative are integrated to address the critical challenges in object tracking. Under this, authors have proposed the sequential execution of both generative as well as discriminative tracker [23]. For fast and accurate performance, simplified PF with fewer particles are used in the appearance model. Hash fingerprints are used in the discriminative tracker to reduce the expensive computation load. Dou et al. [24] have proposed to determine the spatial configuration of each local patch along with the optical flow and Delaunay Triangulation-based inter-frame matching in the appearance model using PF. A discriminative classifier based on a support vector machine (SVM) is used to determine the classification scores for template classification. In [25], the salient-sparse model is used to develop both feature maps as well as to distinguish the target from the background. The update model exploits salient correction for the tracker's stability during heavy occlusions. However, the authors have proposed subspace learning for the adaptive update of the target's template in the generative model [26]. The discriminative classifier is based on weighted confidence values of foreground and background regions for better discrimination of the target in presence of noise. Yi et al. [40] have considered spatial information in the local patches using sparse representation, and weighted local patches using a weighted pooling alignment model. A global discriminate model is used for positive and negative image patches along with confidence score-based update strategy. The authors have proposed a collaborative approach using generative and discriminative models based on coding methods and graph-regularized discriminate algorithms, respectively [41]. Positive and negative templates are classified using the KNN classifier for reliable estimation. Zhao et al. [42] have proposed a hybrid tracking algorithm using sparse coding with PF in the generative model along with two ridge regression models in the discriminative model. Multi-scale sparse coding and ridge regression-based discriminative models are efficient enough to handle scale variations effectively. In sum, hybrid approaches combine the advantages of both strategies not only to maintain the tracker's accuracy during environmental variations but also to enhance discriminative power for tracker stability. Table 2.6 shows the salient features of the representative work under hybrid tracking algorithms.

2.3 DEEP LEARNING-BASED METHODS

Recently, DL-based tracking methods have gained significant attention due to their robust feature representation and learning strategy [43, 44]. In contrast to handcrafted features, deep features can capture the sophisticated multi-level feature relationship efficiently [45, 46]. In particular, deep features-based trackers either fused the deep features extracted from the target or the hierarchical features extracted from the different layers of the deep neural network [47–50]. The details about the deep neural network-based visual tracking algorithms are as follows.

2.3.1 Typical deep learning-based visual tracking methods

Typical DL-based visual tracking methods utilized deep features in the target appearance model to prevent tracking failures. In this direction, Wang et al. [44] have integrated CNN features along with color and HOG features to enhance the tracker's discriminative ability. Color feature is differentiated between the target and the background. HOG feature defines the target-specific properties whereas the CNN feature described the target's generalized properties. A search strategy based on spatial consistency is employed to address occlusion and pose variations. Similarly, Danelljan et al. [45] have proposed to fuse deep RGB and deep optical flow features along with

Table 2.6 Representative work under the hybrid approach

SN	*Reference*	*Sensor*	*Framework*	*Extracted feature*	*Summary*
1.	Dai et al. [23]	Vision	Modified PF and hash tracking	Spatio-temporal contextual features	Generative target model using modified PF with the hash method in the discriminative model for fine-tuning.
2.	Dou et al. [24]	Vision	Sparse representation and Bayesian inference framework	Gray values and HOG	Generative target model using weighted structural local sparse with SVM-based discriminative classifier.
3.	Liu et al. [25]	Vision	Sparse representation	Color, Luminance, and Orientation	Salient-sparse-based generative and collaborative appearance model for discriminative model.
4.	Wang et al. [26]	Vision	Subspace learning	Holistic features	Generative model using histogram-based subspace learning and confidence value-based discriminative model.
5.	Yi et al. [40]	Vision	PF and discriminative online learning	Local histogram and spatial information	Generative model using local histogram and weighted alignment pooling layer. Sparse representation is used in the global discriminant model.
6.	Zhou et al. [41]	Vision	Bayesian framework and graph-regularized discriminant analysis	Holistic features and Local patches	Generative target model using coding method and graph-regularized discriminant analysis, algorithm-based discriminative model.
7.	Zhao et al. [42]	Vision	Sparse coding and ridge regression	CN and HOG	Generative target model using local sparse coding and two ridge regression for the discriminative model.

handcrafted features to improve tracking performance. The strength of the deep feature is evaluated against the handcrafted features to prove their superiority. In [46], the authors have extracted deep features using CNN in the PF framework. Hybrid gravitational search algorithm is used to increase the particle convergence and contribution for final state estimation. However, the authors have utilized CNN for online feature learning along with structural and truncated loss functions for reliable tracking [47]. Temporal sampling based on stochastic gradient descent for the iterative update of the tracker parameters. This update strategy maintains the tracker's performance during heavy occlusions and prevents false detections for efficient results. In summary, deep features incorporated with handcrafted features reduce the impact of noise on the tracker's appearance model. The tracking results achieve high accuracy by capturing the dynamic environmental variations efficiently.

2.3.2 Hierarchical-feature-based visual tracking methods

In this section, we will discuss hierarchical feature-based visual tracking methods. Hierarchical features are those features that are extracted from the different layers of a deep neural network. Lower layers of the deep neural network provide substantial details for target localization whereas higher layers extract the target's semantic and spatial features [48–50]. In [48], the authors have designed a Siamese CNN network for extracting the hierarchical features. Pre-trained Siamese in a spatial-aware network is used for locating the target. The spatial-aware network consists of a spatial transformer and channel attention network to address rotational and scaling challenges. The final confidence response map is generated by calculating the similarity between the target template and the candidate template. However, Zhu et al. [50] have proposed a fully connected Siamese network to compute the similarity between the different frames. Feature maps from the lower layers (fourth layer) are fused with higher-layer feature maps (fifth layer). Lower layers provide the discriminant and rich information for tracking whereas the higher-layer determines the semantic and detailed information in the feature map. Cross-correlation between the fourth layer feature maps is computed in the fully connected network for accurate tracking results. The authors have proposed multi-layer fusion of multi-modal information obtained from the vision sensors and IR sensors [49]. Visible and IR information is processed through two dynamic Siamese networks independently. Features obtained from corresponding layers of the Siamese network are fused to obtain the final response map for target localization. Experimental results infer that the proposed tracker is efficient in comparison to the tracker's exploiting information from the single sensors. DL trackers will be detailed and further elaborated on in Chapter 9. The next section will detail the emergence of tracking algorithms in CF-based trackers.

2.4 CORRELATION FILTER-BASED VISUAL TRACKERS

Recently, CF-based visual trackers have shown excellent performance in terms of speed and accuracy for the target's state estimation. CF-based trackers train the filter in the first frame to model the target appearance model. The peak response map is considered to be the final state computed by the convolution of the filters and samples. However, limitations of the tracking samples restrict the efficiency of these trackers. To address this, CF-based trackers are explored with either context-aware strategy with subspace learning [51–53] or deep features extracted from CNN for computational accuracy [54–56].

2.4.1 Correlation filter-based trackers with context-aware strategy

To enhance the discriminability of the tracker for separating the target from the background, context fusion, and subspace learning constraints were exploited in the tracker's appearance model [52]. Context information is gathered from the background for improved target representation and subspace constraints for better localization accuracy. CF have provided prior information considering the previous frame for target projection in the current frame. The update strategy ensures the accurate update of the tracker with reliable samples to prevent tracking failures. Chen et al. [51] have proposed a weighted adaptive update strategy for the model update in an enhanced structural CF framework. Target's holistic and local part information is exploited to search for the most suitable position for the target. The impact of circular shifting is minimized by enhancing the actual image using the object surrounding the histogram model and Bayes classifier. However, the scale estimation strategy diminishes the computational speed of this tracker. To improve the computational speed of the CF-based trackers, the authors have optimized the filter using the augmented Lagrangian method iteratively [53]. The adaptive group elastic net regularization method is adopted to select the relevant group features to reduce the filter processing load and enhance discriminability. Tracker's performance on multiple public data sets reveals its superiority in comparison to other state-of-the-art. The discussion related to CF-based trackers in deep learning networks is as follows.

2.4.2 Correlation filter-based trackers with deep features

Apart from the handcrafted features, deep features-based CF trackers have shown impressive results because of their discriminative and strong feature representation. In this direction, Nai et al. [54] have extracted high-dimensional CNN features to increase the discriminative power of the tracker. The reliable feature channels are selected adaptively by eliminating noise and redundant features. This not only prevents the overfitting of the CNN features, but also boosts the discriminability of the features. To cater the tracker's drift during complex situations, the temporal regularization term is considered in the tracker's appearance model. On the other hand, the authors proposed a multi-task CF-based tracker in the PF framework [55]. Multiple features are extracted from the intrinsic CNN layers, namely *conv3-4*, *conv4-4*, and *conv5-4*, for each particle. Weighted average response map for every particle is computed for target state estimation. Incremental update strategy using new samples in the current frame is adopted for consistent updates of the tracker. In [56], fine-grain and semantic information for strong feature representation is extracted from the different layers of deep neural networks. Interdependencies among the features extracted from multiple layers are modeled using a multi-task learning scheme. Alternating minimization algorithm is exploited for optimizing the CNN network and CF framework jointly. Tracking results on benchmark data sets prove the robustness of the tracker in complex tracking scenarios. In sum, the strengths of CF and deep features are integrated for strong feature representation to achieve tremendous improvement in tracking outcomes. CF-based trackers will be more elaborated in detail in Chapter 10.

2.5 SUMMARY

In this chapter, we have discussed the various tracking challenges that need to be addressed for practical applications of object tracking. The various categories of the tracker under the traditional framework are discussed to highlight their potential features. Robust trackers based on a probabilistic approach under stochastic framework are discussed due to simple design and architecture. Strategies are adopted to minimize the computational load by reducing the number of processing particles. However, these trackers are not efficient against heavy occlusion and similar background clutters. These limitations are addressed in the deterministic trackers which are based on mean shift and tracking by detection framework. Mean shift-based trackers have a fast convergence rate and provide accurate tracking results. On the other hand, generative-based trackers search for the similarity match between the reference space and target region. These trackers utilize sparse representation in the appearance model and hence, are robust to illumination variations, deformations, and rotational challenges. But multi-view sparse computation is not only time-consuming, but it also amplifies the structural complexity. Discriminative trackers have the potential to achieve significant performance in long-term tracking. These trackers have a discriminative ability to distinguish between the target and background, and achieve superior performance in comparison to generative trackers in presence of background noise and redundant templates. On the other hand, multi-stage trackers reduce computational complexity by estimating the state in two stages. Roughly localized targets in the first stage not only improve the precise state estimation but also minimize computational time. Another category of trackers utilizes collaborative tracking solutions which integrate the advantages and benefits of both approaches. The tracking appearance model is built from a generative approach to improving the template matching process followed by a discriminative classifier. The discriminative classifier discriminates the target from the background during complex background clutters and heavy occlusion. Mostly, traditional tracking approaches utilize handcrafted features extracted from vision sensors or multi-modal sensors in their appearance model. Handcrafted features such as color, texture, gradient, and many more that are inefficient for the high-level and strong feature presentation.

DL-based trackers utilize deep features in their appearance model which are extracted by learning features from CNN layers. Deep features not only address the limitation of handcrafted features, but also develop a strong relationship between the target's features. Deep features integrate the feature information obtained from the consecutive frames for better state estimation. Deep learning trackers are rich in dynamic target information which is able to cater for rotational, translation, and scale variations efficiently. Also, tracking during fast camera movement with small target size can be improved by investigating the hierarchical convolutional features in the appearance model. Deep trackers have shown outstanding performance on standard benchmark video data sets. But the requirement of large training data and specialized hardware limits their applicability to long-term visual tracking applications, and so demands further investigation.

To address the speed and accuracy limitations during long-term tracking, CF has been investigated in visual tracking applications. CF-based trackers consider tracking as a regression problem and exploit the benefits of Fourier transforms for faster computations. Primarily, CF-based trackers utilize either context-aware strategy or deep features in their appearance model. Context-aware features prevent the tracker's drift during similar backgrounds and fast motion. CNN features in the CF framework are more suitable for semantic feature extraction and tracker update. But tracking performance degrades during abrupt scale changes and heavy occlusion in long-term tracking.

To summarize, there is an emergence from traditional approaches to deep neural network-based approaches in visual tracking applications. Each tracking approach has its benefits, strengths, weaknesses, and limitations that need to be explored for better tracking results.

REFERENCES

1. Wu, Y., J. Lim, and M.-H. Yang. Online object tracking: A benchmark. in *Proceedings of the IEEE Conference on Computer Vision and Pattern Recognition*. 2013.
2. Mueller, M., N. Smith, and B. Ghanem. A benchmark and simulator for uav tracking. in *European Conference on Computer Vision*. 2016. Springer.

3. Kumar, A., G.S. Walia, and K. Sharma, Real-time visual tracking via multi-cue based adaptive particle filter framework. *Multimedia Tools and Applications*, 2020. **79**(29): pp. 20639–20663.
4. Narayana, M., H. Nenavath, S. Chavan, and L.K. Rao, Intelligent visual object tracking with particle filter based on modified Grey Wolf optimizer. *Optik*, 2019. **193**: p. 162913.
5. Taylor, L.E., M. Mirdanies, and R.P. Saputra, Optimized object tracking technique using Kalman filter. arXiv preprint arXiv:2103.05467, 2021.
6. Walia, G.S., A. Kumar, A. Saxena, K. Sharma, and K. Singh, Robust object tracking with crow search optimized multi-cue particle filter. *Pattern Analysis and Applications*, 2020. **23**(3): pp. 1439–1455.
7. Dhassi, Y. and A. Aarab, Visual tracking based on adaptive mean shift multiple appearance models. *Pattern Recognition and Image Analysis*, 2018. **28**(3): pp. 439–449.
8. Phadke, G. and R. Velmurugan, Mean LBP and modified fuzzy C-means weighted hybrid feature for illumination invariant mean shift tracking. *Signal, Image and Video Processing*, 2017. **11**(4): pp. 665–672.
9. Rowghanian, V. and K. Ansari-Asl, Object tracking by mean shift and radial basis function neural networks. *Journal of Real-Time Image Processing*, 2018. **15**(4): pp. 799–816.
10. Yu, W., X. Tian, Z. Hou, Y. Zha, and Y. Yang, Multi-scale mean shift tracking. *IET Computer Vision*, 2015. **9**(1): pp. 110–123.
11. Feng, P., C. Xu, Z. Zhao, F. Liu, C. Yuan, T. Wang, and K. Duan, Sparse representation combined with context information for visual tracking. *Neurocomputing*, 2017. **225**: pp. 92–102.
12. Hu, M., Z. Liu, J. Zhang, and G. Zhang, Robust object tracking via multi-cue fusion. *Signal Processing*, 2017. **139**: pp. 86–95.
13. Kang, B., W.-P. Zhu, D. Liang, and M. Chen, Robust visual tracking via nonlocal regularized multi-view sparse representation. *Pattern Recognition*, 2019. **88**: pp. 75–89.
14. Liu, G., Robust visual tracking via smooth manifold kernel sparse learning. *IEEE Transactions on Multimedia*, 2018. **20**(11): pp. 2949–2963.
15. Du, D., H. Qi, L. Wen, Q. Tian, Q. Huang, and S. Lyu, Geometric hypergraph learning for visual tracking. *IEEE Transactions on Cybernetics*, 2017. **47**(12): pp. 4182–4195.
16. Hu, Q., Y. Guo, Z. Lin, W. An, and H. Cheng, Object tracking using multiple features and adaptive model updating. *IEEE Transactions on Instrumentation and Measurement*, 2017. **66**(11): pp. 2882–2897.
17. Lu, R., W. Xu, Y. Zheng, and X. Huang, Visual tracking via probabilistic hypergraph ranking. *IEEE Transactions on Circuits and Systems for Video Technology*, 2017. **27**(4): pp. 866–879.
18. Ma, B., H. Hu, J. Shen, Y. Zhang, L. Shao, and F. Porikli, Robust object tracking by nonlinear learning. *IEEE Transactions on Neural Networks and Learning Systems*, 2017. **29**(10): pp. 4769–4781.
19. Kumar, A., G.S. Walia, and K. Sharma, A novel approach for multi-cue feature fusion for robust object tracking. *Applied Intelligence*, 2020. **50**(10): pp. 3201–3218.

20. Walia, G.S., S. Raza, A. Gupta, R. Asthana, and K. Singh, A novel approach of multi-stage tracking for precise localization of target in video sequences. *Expert Systems with Applications*, 2017. **78**: pp. 208–224.
21. Kumar, A., G.S. Walia, and K. Sharma, Robust object tracking based on adaptive multi-cue feature fusion. *Journal of Electronic Imaging*, 2020. **29**(6): p. 063001.
22. Wu, F., C.M. Vong, and Q. Liu, Tracking objects with partial occlusion by background alignment. *Neurocomputing*, 2020. **402**: pp. 1–13.
23. Dai, M., S. Cheng, and X. He, Hybrid generative–discriminative hash tracking with spatio-temporal contextual cues. *Neural Computing and Applications*, 2018. **29**(2): pp. 389–399.
24. Dou, J., Q. Qin, and Z. Tu, Robust visual tracking based on generative and discriminative model collaboration. *Multimedia Tools and Applications*, 2017. **76**(14): pp. 15839–15866.
25. Liu, Y., F. Yang, C. Zhong, Y. Tao, B. Dai, and M. Yin, Visual tracking via salient feature extraction and sparse collaborative model. *AEU-International Journal of Electronics and Communications*, 2018. **87**: pp. 134–143.
26. Wang, Y., X. Luo, L. Ding, and S. Hu, Multi-task based object tracking via a collaborative model. *Journal of Visual Communication and Image Representation*, 2018. **55**: pp. 698–710.
27. Cai-Xia, M. and Z. Xin-Yan, Object tracking method based on particle filter of adaptive patches combined with multi-features fusion. *Multimedia Tools and Applications*, 2019. **78**(7): pp. 8799–8811.
28. Walia, G.S. and R. Kapoor, Robust object tracking based upon adaptive multi-cue integration for video surveillance. *Multimedia Tools and Applications*, 2016. **75**(23): pp. 15821–15847.
29. Liu, Q., T. de Campos, W. Wang, P. Jackson, and A. Hilton. Person tracking using audio and depth cues. in *Proceedings of the IEEE International Conference on Computer Vision Workshops*. 2015.
30. Fatemeh Razavi, S., H. Sajedi, and M. Ebrahim Shiri, Integration of colour and uniform interlaced derivative patterns for object tracking. *IET Image Processing*, 2016. **10**(5): pp. 381–390.
31. Liu, J. and X. Zhong, An object tracking method based on mean shift algorithm with HSV color space and texture features. *Cluster Computing*, 2019. **22**(3): pp. 6079–6090.
32. Xiao, J., R. Stolkin, Y. Gao, and A. Leonardis, Robust fusion of color and depth data for RGB-D target tracking using adaptive range-invariant depth models and spatio-temporal consistency constraints. *IEEE Transactions on Cybernetics*, 2017. **48**(8): pp. 2485–2499.
33. Wang, Y., X. Luo, L. Ding, and S. Hu, Visual tracking via robust multi-task multi-feature joint sparse representation. *Multimedia Tools and Applications*, 2018. **77**(23): pp. 31447–31467.
34. Li, C., X. Sun, X. Wang, L. Zhang, and J. Tang, Grayscale-thermal object tracking via multi-task laplacian sparse representation. *IEEE Transactions on Systems, Man, and Cybernetics: Systems*, 2017. **47**(4): pp. 673–681.
35. Ma, Z.-A. and Z.-Y. Xiang, Robust object tracking with RGBD-based sparse learning. *Frontiers of Information Technology & Electronic Engineering*, 2017. **18**(7): pp. 989–1001.

36. Li, S., D. Du, L. Wen, M.-C. Chang, and S. Lyu. Hybrid structure hypergraph for online deformable object tracking. in *2017 IEEE International Conference on Image Processing (ICIP)*. 2017. IEEE.
37. Du, D., L. Wen, H. Qi, Q. Huang, Q. Tian, and S. Lyu, Iterative graph seeking for object tracking. *IEEE Transactions on Image Processing*, 2018. **27**(4): pp. 1809–1821.
38. Chen, J.-C. and Y.-H. Lin, Accurate object tracking system by integrating texture and depth cues. *Journal of Electronic Imaging*, 2016. **25**(2): p. 023003.
39. Walia, G.S., H. Ahuja, A. Kumar, N. Bansal, and K. Sharma, Unified graph-based multi-cue feature fusion for robust visual tracking. *IEEE Transactions on Cybernetics*, 2019. **50**(6): pp. 2357–2368.
40. Yi, Y., Y. Cheng, and C. Xu, Visual tracking based on hierarchical framework and sparse representation. *Multimedia Tools and Applications*, 2018. **77**(13): pp. 16267–16289.
41. Zhou, T., Y. Lu, and H. Di, Locality-constrained collaborative model for robust visual tracking. *IEEE Transactions on Circuits and Systems for Video Technology*, 2017. **27**(2): pp. 313–325.
42. Zhao, Z., L. Xiong, Z. Mei, B. Wu, Z. Cui, T. Wang, and Z. Zhao, Robust object tracking based on ridge regression and multi-scale local sparse coding. *Multimedia Tools and Applications*, 2020. **79**(1): pp. 785–804.
43. Kumar, A., Walia, G.S., & Sharma, K. (2020). Recent trends in multicue based visual tracking: A review. *Expert Systems with Applications*, 162, 113711.
44. Wang, J., H. Zhu, S. Yu, and C. Fan, Object tracking using color-feature guided network generalization and tailored feature fusion. *Neurocomputing*, 2017. **238**: pp. 387–398.
45. Danelljan, M., G. Bhat, S. Gladh, F.S. Khan, and M. Felsberg, Deep motion and appearance cues for visual tracking. *Pattern Recognition Letters*, 2019. **124**: pp. 74–81.
46. Kang, K., C. Bae, H.W.F. Yeung, and Y.Y. Chung, A hybrid gravitational search algorithm with swarm intelligence and deep convolutional feature for object tracking optimization. *Applied Soft Computing*, 2018. **66**: pp. 319–329.
47. Li, H., Y. Li, and F. Porikli, Deeptrack: Learning discriminative feature representations online for robust visual tracking. *IEEE Transactions on Image Processing*, 2015. **25**(4): pp. 1834–1848.
48. Li, X., Q. Liu, N. Fan, Z. He, and H. Wang, Hierarchical spatial-aware Siamese network for thermal infrared object tracking. *Knowledge-Based Systems*, 2019. **166**: pp. 71–81.
49. Zhang, X., P. Ye, S. Peng, J. Liu, and G. Xiao, DSiamMFT: An RGB-T fusion tracking method via dynamic Siamese networks using multi-layer feature fusion. *Signal Processing: Image Communication*, 2020. **84**: p. 115756.
50. Zhu, S., Z. Fang, and F. Gao, Hierarchical convolutional features for end-to-end representation-based visual tracking. *Machine Vision and Applications*, 2018. **29**(6): pp. 955–963.
51. Chen, K., W. Tao, and S. Han, Visual object tracking via enhanced structural correlation filter. *Information Sciences*, 2017. **394**: pp. 232–245.
52. Kumar, A., R. Jain, V. A. Devi, & A. Nayyar, (Eds.). *Object Tracking Technology: Trends, Challenges, Impact, and Applications*, 2023. Springer.

53. Xu, T., Z. Feng, X.-J. Wu, and J. Kittler, Adaptive channel selection for robust visual object tracking with discriminative correlation filters. *International Journal of Computer Vision*, 2021. **129**(5): pp. 1359–1375.
54. Nai, K., Z. Li, and H. Wang, Learning channel-aware correlation filters for robust object tracking. *IEEE Transactions on Circuits and Systems for Video Technology*, 2022 **32**(11): pp. 7843–7857.
55. Zhang, T., C. Xu, and M.-H. Yang, Learning multi-task correlation particle filters for visual tracking. *IEEE Transactions on Pattern Analysis and Machine Intelligence*, 2018. **41**(2): pp. 365–378.
56. Zheng, Y., X. Liu, X. Cheng, K. Zhang, Y. Wu, and S. Chen, Multi-task deep dual correlation filters for visual tracking. *IEEE Transactions on Image Processing*, 2020. **29**: pp. 9614–9626.

Chapter 3

Saliency feature extraction for visual tracking

3.1 FEATURE EXTRACTION FOR APPEARANCE MODEL

Object tracking methods extract features, namely, color, texture, edge, thermal, depth, audio, shape, orientation, and many more to keep track of the target in a video stream. It has been well-debated that trackers based on a single feature have shown degraded performance in complex videos. Trackers based on complementary features can address tracking challenges to a great extent [1].

Table 3.1 shows the various extracted features along with their variants and characteristics for object tracking. Features can be extracted utilizing the information either from the vision sensor or a specialized sensor. Vision sensors extract features; specifically color, texture, motion, and gradient. Specialized sensors can be exploited to extract depth, thermal, and audio information. Color features can be extracted either as RGB through color histogram (CH) or hue saturation value (HSV) methods. This feature is easy to extract and requires less processing in terms of computations. But it has limited capability to track the target during the night and in low-resolution videos. Also, it is inefficient to address abrupt illumination variations and background clutters. The texture feature can handle these challenges efficiently as it checks for repetitive information after certain intervals. But requires processing methods for its extraction and is not able to address scale, rotational, and camera translation variations. Generally, LBP (linear binary pattern) [2], CLTP (completed linear ternary pattern) [3] and mLBP (mean linear binary pattern) [4] are a few texture feature extraction methods exploited for tracking objects.

Motion features can handle fast motion and motion blur challenges but inefficient to handle rotation and scale variations. Optical flow [26] is commonly used for extracting motion features. This feature can handle heavy occlusion but is not able to address the target deformations. Gradient features can address shape deformations due to pose variations but are not able to address similar backgrounds. This feature can be computed using either HOG (histogram of gradient) [27] or PHOG (pyramid of histogram of gradient) [28].

DOI: 10.1201/9781003456322-3

Table 3.1 Descriptions of feature extraction for tracking, their variants, and characteristics

SN	*References*	*Sensors*	*Extracted features*	*Variants*	*Summary*
1.	[5–7]	Vision	Color	CH, HSV	Easy to extract and simple to compute. Inefficient to track during abrupt illumination variations and in low-resolution videos.
2.	[4, 8, 9]		Texture	LBP, CLTP, mLBP	Execute in real-time, multiple variants are available to ease the tracking task. Inefficient for rotational variations and abrupt camera movement.
3.	[10–12]		Gradient	HOG, PHOG	Invariant to scale and pose variations. Degrades during object deformations.
4.	[13, 14]		Motion	Optical flow	Consider information from the consecutive frames and can handle rotational variations. Neglects the shape and gradient information.
5.	[15, 16]	Specialized	Thermal	–	Insensitive to abrupt illumination change. Inefficient to differentiate between targets with a similar thermal profile.
6.	[17, 18]		Depth	–	Suitable to address heavy occlusion. Failed to track a fast-moving target.
7.	[19, 20]		Audio	TDOA	Efficient to out-of-view and small-size target challenges. Not able to discriminate between the foreground and background audio information.
8.	[21]	Deep Neural Network	Deep RGB	CNN in RGB network	CNN trained by an RGB network to obtain a robust feature invariant to change in target appearance.
9.	[22]		Deep motion	CNN in motion network	Capture the target's high-level motion information efficiently.
10.	[23–25]		Hierarchical	Hyper-Siamese network, Fully CNN, Spatial-aware Siamese CNN	Hyper-features either from the deepest layers of hierarchical layers were used for precise target representation.

Further, the thermal feature from the thermal sensor extracts the thermal profile of the target from the scene [15]. It is invariant to illumination variations, and pose variations can enhance the tracker's performance during the night. However, trackers based on thermal profile alone are insufficient to discriminate between similar thermal profile targets, and their performance is impacted by rapid changes in temperature. Depth feature extracts either from depth camera or Kinect sensors [17]. Depth information is efficient to address heavy occlusion, but it changes rapidly during target motion. Depth data has no texture information and hence, is not suitable for handling rotational challenges either due to target or camera. Audio information can be integrated with depth information to address its limitations [19]. Audio features can efficiently handle the out-of-view and occlusion challenges, but trackers based on an audio feature only are not able to differentiate between the foreground and background information. Also, audio information captured from a distant sensor is not reliable for efficient tracking results.

In another line of research, deep neural networks are also used for the extraction of deep features. CNN networks are trained with RGB information and motion information to extract the deep RGB and deep motion [21]. Deep RGB can be computed from the first layer and deep layer of an RGB network. The deep motion feature is computed by training a CNN with a motion network. CNN model is trained with optical flow images to generate a high-level motion feature. This feature can capture the dynamic variations in the scene essential for a stable tracker. On the other hand, hierarchical features are also extracted from the deepest and the higher layers of CNN network [23–25]. These features are extracted from fully connected CNN, and hyper-Siamese for providing end-to-end tracking solutions. Hierarchical features are not only discriminative but also invariant to target deformation. These features are rich in the target's semantic and structural information essential for better tracking results. The next section will detail the various feature extraction methodologies.

3.2 HANDCRAFTED FEATURES

Handcrafted features capture low-level information which is easy to extract and requires less processing power for computation. These features can be extracted either from vision sensors or specialized sensors. Complementary features are integrated to prevent the tracker's drift during complex environmental variations. The elaborative details about the various feature extraction methods in visual tracking are as follows.

3.2.1 Feature extraction from vision sensors

Vision sensors can be used to extract the color, texture, gradient, motion, and shape information from the target. The details for feature extraction are as follows.

3.2.1.1 *Color feature*

Color features can be extracted from CH and HSV models. CH can be used to extract the RGB information. Color histogram (C_b) can be computed using Eq. (3.1).

$$C_b = \delta \sum_{p=1}^{N} B(x_p, y_p), b = 1,2,3 \ldots N_b \tag{3.1}$$

where $B(.)$ is the binning function that assigns pixel (x_p, y_p) to one of the N_b histogram bin. δ is the normalizing factor which is calculated using $\sum_{b=1}^{N_b} C_b = 1$.

Alternatively, the HSV model can also be used for color feature extraction. It is based on the capacity of human eye perception. HSV are represented as cylindrical geometries in which hue is used as angular dimension. Primarily, red color at (0°), green at (120°), blue at (240°), and finally back to red at (360°). In geometric representation, neutral, achromatic, or gray values comprise the central vertical axis ranging from black at value 0 to white at value 1. Each weight of the HSV is quantized to obtain the color values.

3.2.1.2 *Texture feature*

Generally, texture features can be computed efficiently using rotational invariant LBP, CLTP, and mLBP. LBP considering the neighboring (k_{th}) pixel grey value ($\mathcal{G}_k$) and the central pixel grey value ($\mathcal{G}_c$) with radius (r) and equally spaced neighboring pixels (n_p) can be computed using Eq. (3.2).

$$\text{LBP}_{r,n_p} = \begin{cases} \sum_{k=1}^{n_p-1} T(\mathcal{G}_k - \mathcal{G}_c) & , \quad \text{if } U(\text{LBP}_{r,n_p}) \le 2 \\ p+1 & , \quad \text{otherwise} \end{cases} \tag{3.2}$$

where T (.) is the step function. $U(\text{LBP}_{r,n_p})$ measures the uniform pattern bitwise transition labels in the image labels [2].

Also, CLTP [9] quantifies the intensity difference between the central pixel and the (k_{th}) pixel in the neighborhood with radius (r). Difference (D_K) can be calculated as $(D_k = \mathcal{G}_k - \mathcal{G}_c)$ and represented using Eq. (3.3).

$$D_k = \mu_p * \mathcal{M}_p \tag{3.3}$$

where $\mathcal{M}_p = |D_k|$. For CLTP, μ_p can be defined as Eq. (3.4).

$$\mu_p = \begin{cases} 1 & D_k \geq \varphi \\ 0 & |D_k| < \varphi \\ -1 & D_k \leq \varphi \end{cases} \tag{3.4}$$

where φ is the threshold value. The CLTP can be computed for the three operators namely, magnitude, positive, and negative using Eqs. (3.5)–(3.7), respectively.

$$\text{CLTP}^m = \sum_{k=0}^{k-1} \mu_1(\mathcal{M}_p, \varphi), \mu_1(\gamma, \varepsilon) = \begin{cases} 1, & \gamma \geq \varepsilon \\ 0, & \text{otherwise} \end{cases} \tag{3.5}$$

$$\text{CLTP}^+ = \sum_{k=0}^{k-1} \mu_2(\mathcal{M}_p, \varphi) 2^{\mu}, \mu_1(\gamma, \varphi) = \begin{cases} 1, & \gamma \geq \varphi \\ 0, & \text{otherwise} \end{cases} \tag{3.6}$$

$$\text{CLTP}^- = \sum_{k=0}^{k-1} \mu_2(\mathcal{M}_p, \varphi) 2^{\mu}, \mu_1(\gamma, \varphi) = \begin{cases} 1, & \gamma \geq -\varphi \\ 0, & \text{otherwise} \end{cases} \tag{3.7}$$

where ε is adaptively determined using the mean value of $\mathcal{M}_p$.

Generally, LBP is robust to illumination variations but sometimes its performance degrades when there is a drastic change in contrast and lighting in consecutive frames. To address this, modified mLBP was proposed by [4]. For this, the relative difference of k_{th} pixel grey value $(\mathcal{G}_k)$ from the mean (Σ_k) is calculated using Eq. (3.8).

$$\mathcal{G}_k = \mathcal{G}_c - \Sigma_k \tag{3.8}$$

where $\Sigma_k = \frac{\sum_{i=0}^{n_p-1} \mathcal{G}_c(i)}{n_p}$ is the mean of the gray values of equally spaced neighboring pixels (n_p) with radius (r) from c_{th} pixel. Pixel gray values $(\mathcal{G}_k)$ are converted into the binary form using Eq. (3.9).

$$\beta(k) = \begin{cases} 1, & \text{if } \mathcal{G}_k \geq 0, \\ 0, & \mathcal{G}_k < 0 \end{cases} \tag{3.9}$$

where $\beta(k)$ is the binary image representing the binary pattern of c_{th} pixel and its neighborhood. The modified gray value $\mathcal{G}_c'$ of the c_{th} pixel is calculated using Eq. (3.10).

$$\mathcal{G}_c' = \sum_{n_p=0}^{n_p-1} \beta(k) \times 2^{n_p} \tag{3.10}$$

Finally, modified LBP (mLBP) can be computed using Eq. (3.11).

$$m\text{LBP} = \delta' \sum_{i=1}^{n} \rho\left[\mathcal{G}_c' - u\right] \tag{3.11}$$

where δ' is the normalizing factor. n and u are the number of pixel locations and texture bins, respectively.

3.2.1.3 Gradient feature

Gradient feature is used to extract the intensity along the edges. Primarily, HOG and PHOG are used to extract the target's edge information. For this, HOG computes the gradient intensities distribution in both planes; horizontal and vertical. The intensity distribution is obtained by filtering the image with two kernels i.e. $[-1, 0, 1]$ & $[-1, 0, 1]'$. Magnitude (M_k) and orientation for k_{th} pixel can be calculated using Eq. (3.12) and (3.13), respectively.

$$M_k = \sqrt{\left((\mathcal{G}_h)^2 + (\mathcal{G}_v)^2\right)^2} \tag{3.12}$$

$$\theta_k = \tan^{-1}\left(\frac{\mathcal{G}_h}{\mathcal{G}_v}\right) \tag{3.13}$$

where $\mathcal{G}_h$ and $\mathcal{G}_v$ are gradient intensities for k_{th} pixel in horizontal and vertical directions, respectively. The region of interest (ROI) in the image is divided into small rectangular cells. Each pixel in a particular cell casts a weighted vote for its orientation histogram as well as for neighboring pixels. Then, the histogram for each cell in the bin (α) can be calculated using Eq. (3.14).

$$\text{H}(\alpha) = \sum_{i=1}^{T_p} \left(M_{k,i}\, \delta_k\left(\theta_{k,i}' - \alpha\right)\right) \tag{3.14}$$

where θ_k' is quantized orientation calculated from θ_k. $\delta_k(.)$ is the Kronecker delta function and T_p is the total number of pixels in a particular cell. L2-norm function is used to obtain illumination variation and contrast invariant gradient values. The histogram values are normalized using Eq. (3.15).

$$\mathrm{H}_n(\alpha) = \frac{\mathrm{H}_n(\alpha)}{\sqrt{\sum_{i=1}^{c \times c \times N_b} \mathrm{H}_n(i)^2 + \kappa^2}} \tag{3.15}$$

where c is the number of cells, N_b is the number of bins per cell, and κ is the regularization parameter.

Further, gradient values can also be computed using the PHOG approach. PHOG can be used to extract the shape edge orientation and spatial information of the edges. For this, ROI in the image is extracted and the gradient intensity value (H) for k_{th} pixel (x_k, y_k) is calculated using Eq. (3.16).

$$\mathrm{H}(x_k, y_k) = \sqrt{\left(\mathcal{G}(x_k, y_{k+1}) - \mathcal{G}(x_k, y_{k-1})\right)^2 + \left(\mathcal{G}(x_{k+1}, y_k) - \mathcal{G}(x_{k-1}, y_k)\right)^2} \tag{3.16}$$

The orientation of the image is divided into bins and orientation can be calculated using Eq. (3.17).

$$\theta = \tan^{-1}\left(\frac{\mathcal{G}(x_k, y_{k+1}) - \mathcal{G}(x_k, y_{k-1})}{\mathcal{G}(x_{k+1}, y_k) - \mathcal{G}(x_{k-1}, y_k)}\right) \tag{3.17}$$

To extract the spatial shape, the image ROI is segmented into regions of low resolution. These regions are concatenated to obtain the final PHOG $(\dot{\mathrm{H}})$ using Eq. (3.18).

$$\dot{\mathrm{H}}(\alpha) = \delta \sum_{i=1}^{T_p} \left(\mathrm{H}(x_k, y_k) \times \delta_k\left(s(x_k, y_k) - \alpha\right)\right) \tag{3.18}$$

where δ is the normalizing constant, $\delta_k(.)$ is the Kronecker delta function, T_p is the total number of pixels in a particular bin, $s(.)$ maps pixels to histogram bins (α).

3.2.1.4 Motion feature

The motion descriptor of the target can be computed by calculating the optical flow using the Horn-Schunck method [26]. Optical flow determines the rate of change of pixel intensity in the consecutive frames considering that there is no change in the brightness level. The pixel intensity $I(x, y, t)$ for pixel (x, y) is depicted using Eq. (3.19).

$$I(x,y,t) = I_x \upsilon_x + I_y \upsilon_y + I_t \tag{3.19}$$

where $\upsilon_x = \frac{dx}{dt}$ and $\upsilon_y = \frac{dy}{dt}$ are the change in pixel displacement w.r.t to change in time t in the x and y directions, respectively. Optical flow velocity vector can be denoted as $\vec{\upsilon} = (\upsilon_x, \upsilon_y) = (dx, dy)$ Smoothness constraint is considered to minimize the magnitude of the optical flow components. The error term for the same is calculated using Eq. (3.20).

$$\xi^2 = \left(\frac{\partial \upsilon_x}{\partial x}\right)^2 + \left(\frac{\partial \upsilon_x}{\partial y}\right)^2 + \left(\frac{\partial \upsilon_y}{\partial x}\right)^2 + \left(\frac{\partial \upsilon_y}{\partial x}\right)^2 \tag{3.20}$$

where $\frac{\partial \upsilon_x}{\partial x}$ and $\frac{\partial \upsilon_x}{\partial y}$ are change in υ_x in x and y directions, respectively. Similarly, $\frac{\partial \upsilon_y}{\partial x}$ and $\frac{\partial \upsilon_y}{\partial x}$ are changed in υ_y and in x and y directions, respectively. The total error is minimized and can be depicted using Eq. (3.21).

$$\varepsilon_T^2 = \left(I(x,y,t)\right) + \xi^2 \left(\left(\frac{\partial \upsilon_x}{\partial x}\right)^2 + \left(\frac{\partial \upsilon_x}{\partial y}\right)^2 + \left(\frac{\partial \upsilon_y}{\partial x}\right)^2 + \left(\frac{\partial \upsilon_y}{\partial x}\right)^2 \right) dx\, dy \tag{3.21}$$

where the second term is the smoothness constraint and ξ^2 is a constant known as the weighting factor. Using Eq. (3.21) optical flow vector $\vec{\upsilon} = (u, v)$ can be computed by minimizing the error term ε_T.

3.2.2 Feature extraction from specialized sensors

Apart from vision cameras, specialized sensors are also used to extract powerful features for robust tracking solutions. A target's depth information from the Kinect sensors, thermal profile from thermal sensors, and discriminating audio information from the audio sensors can be captured to minimize tracking failures.

3.2.2.1 Depth feature

Depth information can be extracted from Kinect sensors, lidar, and PrimeSense. Sample frames from the RGBD Princeton tracking data set (PTD) [29] are depicted in Figure 3.1. Target's depth information is invariant to change in illumination and useful for 3D to 2D real-world representation. In an image, depth data considers the pixel distance between the

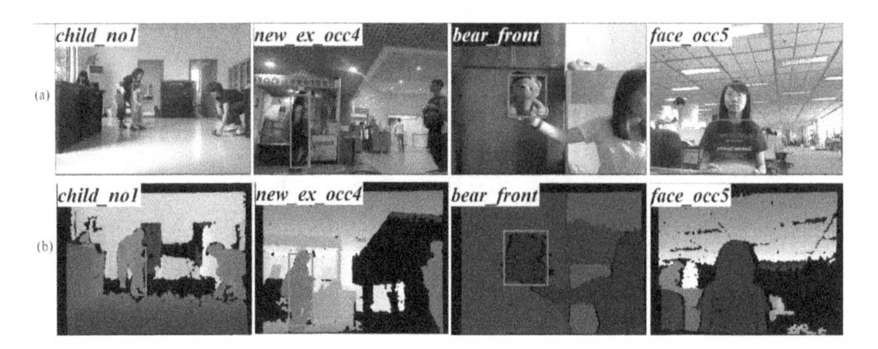

Figure 3.1 Sample frames from PTD [29] data sets representing (a) RGB data (b) depth data. Each video frame is labeled with the video sequence name in the left corner.

target and the camera. This information is more robust and discriminative in comparison to intensity data which is extracted by most of the visual features. So, depth information has been integrated with RGB data to address the occlusion efficiently [17, 18]. In comparison to RGB, depth information changes very rapidly during the target movement. Hence, depth data is primarily used to handle the target re-detection after heavy occlusion and out-of-view challenge [30].

3.2.2.2 *Thermal feature*

Thermal information about the target is captured using either the thermal camera or IR camera. The thermal camera captures the IR radiation emitted by the target with above zero absolute temperature. It has been determined that the human thermal profile is more distinguishable in comparison to other objects [31]. In addition, thermal distribution is invariant to changes in illumination. Hence, thermal data is ideal for tracking objects handling complex environmental conditions. But thermal sensors suffer from the Halo effect. This effect introduces a black ring around a bright object and produces a misleadingly low temperature which leads to tracking failure. Also, targets with similar thermal distribution are hard to discriminate during background clutter and occlusion. For this, thermal data can be complemented with visual data for efficient tracking results [15, 32]. Figure 3.2 illustrates the sample frames for the thermal images and the corresponding RGB images from the OCU color and thermal data set [33].

3.2.2.3 *Audio feature*

Audio feature is extracted from a moving target producing a voice signal for tracking [19, 20]. Time delay of audio arrival (TDOA) is an effective method used to detect the reliable direction of sound source for tracking [34]. The

Figure 3.2 Sample frame from OCU color and thermal data set (a) RGB data (b) Thermal profile data.

normalized cross power-spectrum $\omega_p(t, f)$ centered at time t and frequency f between two spectra $\mathcal{R}_{p1}(t,f)$ and $\mathcal{R}_{p2}(t,f)$ of audio signals from p_{th} microphone pair (p_1, p_2) can be depicted using Eq. (3.22).

$$\omega_p(t,f) = \frac{\mathcal{R}_{p1}(t,f)\mathcal{R}^*_{p2}(t,f)}{|\mathcal{R}_{p1}(t,f)||\mathcal{R}^*_{p2}(t,f)|}, \tag{3.22}$$

where $\mathcal{R}^*_{p2}(t,f)$ is the complex conjugate computed from $\mathcal{R}_{p2}(t,f)$. The coherence measure M_p representing the similarity between two segments for time lag ρ at p_{th} pair microphone can be computed using Eq. (3.23).

$$\mathrm{M}_p(t,\rho) = \int_{-\infty}^{+\infty} \omega_p(t,f)e^{j2\pi f\rho}\, df \tag{3.23}$$

If i is a generic 3D point, then 3D acoustic map $A(t, i)$ for P number of microphone pairs can be computed using Eq. (3.24).

$$A(t,i) = \frac{1}{P}\sum_{p=0}^{P-1} M_p(t, v_p(P)) \tag{3.24}$$

where $\nu_p(P)$ is ideal TDOA at 3D position P w.r.t to p_{th} pair microphone. Finally, the sound source location can be computed using Eq. (3.25).

$$P_t^{\alpha} = \arg\max_{i \epsilon P} \mathrm{A}(t,i) \tag{3.25}$$

where i represents a single point under P set of points of the 3D grid under analysis. It has been observed that the audio feature can only be utilized for tracking when the person is speaking. To expand the functionality and utility of audio features, depth, and visual cues have been integrated with it [19, 20]. Apart from these handcrafted features, more powerful features are extracted from deep neural networks for reliable tracking solutions. The next section will detail the various deep learning-based feature extraction methods.

3.3 DEEP LEARNING FOR FEATURE EXTRACTION

Deep neural networks such as CNN [21], residual networks [35], and LSTM residual networks [36] can be used for more robust features suitable for robust tracking solutions. Deep features are rich in the target's semantic information and have more discriminative power in comparison to handcrafted features. These features can be obtained either by training a deep learning network or from different layers of the network.

3.3.1 Deep features extraction

Deep features such as deep RGB and deep motion can be extracted by training a CNN [21]. Siamese CNN networks are trained to learn local spatio-temporal features for a strong target appearance model [37]. These local features provide the distinguishability of the target from the other objects in the scene. In [38], authors have extracted feature histograms from the features trained by CNN. The so-obtained feature histograms are not only invariant to covariate shifts but also adapt to environmental variations. The histogram of CNN features is tracking-oriented suitably for effective tracking results. However, interdependent channel-wise features are extracted by channel attention using residual learning [35]. Also, the spatial attention module is exploited to capture the target's contextual information. These features are more meaningful and enriched in terms of the discriminative ability of the tracker. To obtain more powerful deep features in terms of spatio-temporal attention information, residual long-short-term memory (RLSTM) networks are used [36]. These features have improved the tracking performance in complex videos. Rule-based RLSTM learning is used to make sure the tracker training is done using reliable features only. Deep learning-based trackers also extract the features from different layers of a deep neural network. The details of the tracker under this category are as follows.

3.3.2 Hierarchical feature extraction

Deep features require a lot of training data and failed to address the background clutter challenge efficiently. To address this, Hierarchical features are proposed to be extracted from different layers of the DL network [23–25].

In [23], feature information from the lower two layers is captured in a Siamese fully connected CNN. The cross-correlation between the feature maps is extracted from the fourth and fifth layer of CNN to obtain a final robust feature map. The lower layers of feature maps consist of the target's spatial information, which is essential for its precise localization. However, the authors proposed to extract the target's semantic information from the higher layers and localization details from the lower layers of an attention Siamese network [24]. Adaptive attention weights are determined for generating the feature maps from both layers. Cross-correlation between the middle layers and the last layer feature map is computed based on the attention weights. Attention weights are responsible for making the tracker adaptive to environmental variations. Zhang et al. [25] have proposed to extract the semantic and spatial features from different CNN layers for improving tracking results. The high-level and low-level enriched target information is extracted from the feature maps of *Conv1–2*, *Conv2–2*, *Conv3–4*, *Conv4–4*, and *Conv5–4* layers. Feature maps are obtained from *Conv1–2* and *Conv2–2* layers utilized for spatial details and had abundant target texture and edge information, whereas feature maps obtained from *Conv3–4*, *Conv4–4*, and *Conv5–4* layers are rich in semantic details and catered for the changing environmental situations. In [39], authors have utilized color as a low-level feature and semantic information from the layers of a CNN network to generate hierarchical-feature maps. The score maps for these features are fused to obtain effective tracking accuracy. The next section will detail the various multi-feature techniques.

3.4 MULTI-FEATURE FUSION FOR EFFICIENT TRACKING

Analysis indicates that complementary features are superior to addressing tracking challenges. Complementary features are those features in which one feature will contribute to the performance of another if its performance degrades during tracking variations. For efficient tracking results, these features need to be fused to obtain a final robust feature that is invariant to tracking challenges. In general, tracking features are fused using either score-level fusion or feature-level fusion.

Score-level fusion determines the classifier scores from the features and fused them to obtain a robust feature for segmenting the target from the background [5, 7, 11, 35]. Along this line, authors have adaptively fused three feature scores, namely color, LBP, and PHOG, exploiting rank-based fusion techniques in a PF framework [5]. The fusion techniques ensure the automatic boosting of important particles and suppression of affected particles to create clear decision boundaries for the classifier scores. The outlier detection mechanism based on classifier scores is used to separate the important particles from the affected particles for the final state estimation.

Similarly, Walia et al. [7] have fused two feature scores such as color and texture based on fuzzy inference rules. The outlier detection method based on the Markov process is utilized to separate the important particles from the affected particles efficiently. This method has improved the classifier scores classification to improve the tracker's state estimation. In [11], the authors have proposed a two-stage process from coarse to fine estimation. Borda count ranking fusion is used to integrate the likelihoods of three features namely, color, LBP, and PHOG. Also, the reliability of each feature along with feature likelihoods is fused to assign the final weight to the fragments for target localization. Authors have utilized Siamese networks for target tracking [35]. Features are extracted from the CNN, and score maps are generated by computing the cross-correlation between the features. Finally, the features are weighted in a context-aware attention network to obtain the spatial information and channel dimensions. In summary, score-level fusion integrates the feature score or classifier score to localize the target. However, the final feature performance may be impacted by the trained classifier scores.

It has been evident that feature-level fusion generates more robust and discriminative unified features than score-level fusion [40]. Feature-level fusion generates reliable features invariant to environment variations. In this direction, authors have computed the robust unified feature by integrating the features vectors [8]. Features vectors, namely LBP and HOG, are unified by computing the Euclidean distance between the feature vectors for each fragment. Transductive reliability for the features is calculated to ensure the tracker adapts to changing environmental conditions. Also, authors have proposed a modified cross-diffusion mechanism for fusing features namely, intensity, texture, and HOG [13]. The cross-diffusion mechanism ensures to integrate the salient features for generating the unified feature vector. Sparse and dense similarity feature graphs are fused using recursive normalization to create the precise decision boundary for precise target localization. In summary, feature-level fusion generates reliable and robust unified features suitable for efficient tracking. But processing the final feature map with integrated information will be computationally complex due to the length of the individual feature vector.

In another line of research, score-level and feature-level fusion are also explored to fuse deep features and hyper-features [41–44]. In ref. [41], authors have integrated the handcrafted features and deep CNN features linearly. Handcrafted features, namely color and HOG, are weighted and fused with deep features extracted from *conv5-4* and *conv4-4* layers of a VGGNet. To improve the discriminability of the features, features are weighted adaptively, i.e. handcrafted features are assigned a lower weight in comparison to deep features. The obtained final feature is optimized using Gaussian constrained optimization method to ensure the adaptability of the tracker. To capture the spatial variations, authors have proposed a multi-feature fusion technique to fuse features namely, CN, intensity, HOG, and saliency in a discriminative correlation filter [42]. The response map for each feature is combined using element-wise addition to produce the final precise feature map by considering the peak response of the features. Spatial variations and contextual information for the target are also utilized to enhance the tracker's appearance model. However, the authors have generated a robust response map by combining individual features' response maps using a weighted averaging scheme [43]. A penalty item is employed to assign a large weight to better response map features and a lower weight to low response map features. Zhao et al. [44] have proposed multi-level feature fusion obtained from shallow and deep layers of a CNN. Layer-wise similarity considering the current frame and the previous frame is extracted to identify the weak features to enhance the robustness of the tracker. Layer-wise features obtained from shallow and deep layers are weighted, normalized, and regularized to obtain a robust appearance model. Table 3.2 shows the description of representative work exploiting either the score-level fusion or feature-level fusion.

Table 3.2 Representative work with their extracted features and the fusion technique

SN	*References*	*Extracted features*	*Fusion Technique*	*Summary*
1.	Phadke and Velmurugan [4]	mLBP and modified fuzzy c-means weighted CH	Mean shift-based hybrid feature vector fusion	Utilized robust modified color and texture features invariant to illumination and scale variations.
2.	Kumar et al. [5]	Color, LBP and PHOG	Non-linear rank-based adaptive fusion of particles weight	Computed context cue reliability for the adaptive fusion of particle weights.
3.	Walia et al. [7]	Color and texture	Fuzzy inference rule-based adaptive fusion of particles weight	Determined the affected particles using Markov-based outlier mechanism to improve the tracker's efficiency.
4.	Kumar et al. [8]	LBP and HOG	Euclidean distance-based adaptive feature fusion	Computed transductive reliability for adaptive feature fusion.
5.	Dou et al. [9]	Corrected Background Weighted Histogram, CLTP, and HOG	Important weighted fusion of particles	Adaptive fusion of particle weight along with dynamic adjustment of model probability.
6.	Walia et al. [11]	Color, LBP and PHOG	Borda count rank-based adaptive fusion	Adaptive fusion of feature fragments in a two-stage estimation model.
7.	Cai-Xia et al. [12]	CH and HOG	Weighted fusion of features sub-patches	Particle sub-patch weight adaptively adjusted utilizing particle sub-space information.

(Continued)

Table 3.2 (Continued)

SN	*References*	*Extracted features*	*Fusion Technique*	*Summary*
8.	Walia et al. [13]	Intensity, texture, and HOG	Cross-diffusion based adaptive feature fusion	Sparse and dense graph-based feature fusion with recursive normalization.
9.	Xiao et al. [18]	Color and depth	Extended clustered decision tree-based feature fusion	Depth data is integrated with visual data to prevent false detection during heavy occlusion.
10.	Zhang et al. [35]	Deep features	Weighted sum-based fusion of deep features	Attention-weighted feature fusion with residual learning to prevent tracker's degradation during tracking challenges.
11.	Chen et al. [41]	Color, HOG, and hierarchical features	Weighted adaptive feature fusion	Final feature response map optimized to obtain robust target appearance model.
12.	Elayaperumal & Hoon Joo [42]	CN, intensity, HOG, and saliency	Element-wise addition-based feature fusion	Utilized the foreground and background information based on contextual information.
13.	Yuan et al. [43]	Edge, color, and intensity	Weighted averaging-based feature fusion	Incorporated scale detection method to address the scale variations effectively.
14.	Zhao et al. [44]	Hierarchical features	Multi-level adaptive feature fusion	The fusion mechanism managed the granularity from the shallow to deep features.

3.5 SUMMARY

In this chapter, we have discussed various feature extraction methodologies for developing a robust appearance model. Handcrafted features extracted either from vision cameras or specialized sensors. Visual features such as color, motion, texture, edge, and gradient are considered low-level features. These features are computationally easy to extract and require less processing power. Target's feature information such as depth information, thermal profile, and audio data extracted from specialized sensors can be integrated with visual information to extend its efficiency and effectiveness. Primarily, depth information is integrated with visual features to address the tracking failures during extensive occlusion scenarios. The thermal profile of the target is utilized with the visual data to extend its operations during the night. Also, this potential information is effective to address the heavy background clutters. Audio information has the potential to address rotational as well as out-of-view challenges. Trackers based on handcrafted features are computationally efficient but failed to address the long-term tracking challenges.

To address the limitations of handcrafted features, deep learning-based trackers have presented effective tracking solutions. Deep learning-based trackers have integrated deep features with handcrafted features to develop a strong appearance model. Also, the hierarchical features which are extracted from the shallow and deep layers of a deep neural network are integrated to improve tracking performance. Deep learning-based trackers have shown superior performance in terms of tracking accuracy in comparison to handcrafted-based trackers. But DL-based trackers require huge training data for the analysis. A lot of computational efforts in terms of specialized hardware and processing time are required by these trackers for tracking. These requirements restrict the real-time realization of these trackers to the tracking applications.

It has been well acknowledged that complementary features have the potential to develop a robust appearance model invariant to tracking failures. To develop a strong tracker, these features are integrated using either score-level fusion or feature-level fusion. Score-level fusion utilizes the classifier scores to segment the target from the background. Score-level fusion is computationally simple but introduces the classifier error in the final outcome. On the other hand, feature-level fusion generates unified feature vector invariant to environmental variations. However, processing such a high-dimensional feature vector requires a lot of computational effort and may impact the speed of the tracker.

REFERENCES

1. Kumar, A., G.S. Walia, and K. Sharma, Recent trends in multicue based visual tracking: A review. *Expert Systems with Applications*, 2020. **162**: p. 113711.
2. Ojala, T., M. Pietikainen, and T. Maenpaa, Multiresolution gray-scale and rotation invariant texture classification with local binary patterns. *IEEE Transactions on Pattern Analysis and Machine Intelligence*, 2002. **24**(7): pp. 971–987.
3. Guo, Z., L. Zhang, and D. Zhang, A completed modeling of local binary pattern operator for texture classification. *IEEE Transactions on Image Processing*, 2010. **19**(6): pp. 1657–1663.
4. Phadke, G. and R. Velmurugan, Mean LBP and modified fuzzy C-means weighted hybrid feature for illumination invariant mean-shift tracking. *Signal, Image and Video Processing*, 2017. **11**(4): pp. 665–672.
5. Kumar, A., G.S. Walia, and K. Sharma, Real-time visual tracking via multi-cue-based adaptive particle filter framework. *Multimedia Tools and Applications*, 2020. **79**(29): pp. 20639–20663.
6. Liu, J. and X. Zhong, An object tracking method based on Mean Shift algorithm with HSV color space and texture features. *Cluster Computing*, 2019. **22**(3): pp. 6079–6090.

7. Walia, G.S., A. Kumar, A. Saxena, K. Sharma, and K. Singh, Robust object tracking with crow search optimized multi-cue particle filter. *Pattern Analysis and Applications*, 2020. **23**(3): pp. 1439–1455.
8. Kumar, A., G.S. Walia, and K. Sharma, A novel approach for multi-cue feature fusion for robust object tracking. *Applied Intelligence*, 2020. **50**(10): pp. 3201–3218.
9. Dou, J. and J. Li, Robust visual tracking based on interactive multiple model particle filter by integrating multiple cues. *Neurocomputing*, 2014. **135**: pp. 118–129.
10. Du, D., L. Wen, H. Qi, Q. Huang, Q. Tian, and S. Lyu, Iterative graph seeking for object tracking. *IEEE Transactions on Image Processing*, 2018. **27**(4): pp. 1809–1821.
11. Walia, G.S., S. Raza, A. Gupta, R. Asthana, and K. Singh, A novel approach of multi-stage tracking for precise localization of target in video sequences. *Expert Systems with Applications*, 2017. **78**: pp. 208–224.
12. Cai-Xia, M. and Z. Xin-Yan, Object tracking method based on particle filter of adaptive patches combined with multi-features fusion. *Multimedia Tools and Applications*, 2019. **78**(7): pp. 8799–8811.
13. Walia, G.S., H. Ahuja, A. Kumar, N. Bansal, and K. Sharma, Unified graph-based multicue feature fusion for robust visual tracking. *IEEE Transactions on Cybernetics*, 2019. **50**(6): pp. 2357–2368.
14. Kumar, A., G.S. Walia, and K. Sharma, Robust object tracking based on adaptive multicue feature fusion. *Journal of Electronic Imaging*, 2020. **29**(6): p. 063001.
15. Li, C., X. Sun, X. Wang, L. Zhang, and J. Tang, Grayscale-thermal object tracking via multitask laplacian sparse representation. *IEEE Transactions on Systems, Man, and Cybernetics: Systems*, 2017. **47**(4): pp. 673–681.
16. Walia, G.S. and R. Kapoor, Robust object tracking based upon adaptive multi-cue integration for video surveillance. *Multimedia Tools and Applications*, 2016. **75**(23): pp. 15821–15847.
17. Ma, Z.-A. and Z.-Y. Xiang, Robust object tracking with RGBD-based sparse learning. *Frontiers of Information Technology & Electronic Engineering*, 2017. **18**(7): pp. 989–1001.
18. Xiao, J., R. Stolkin, Y. Gao, and A. Leonardis, Robust fusion of color and depth data for RGB-D target tracking using adaptive range-invariant depth models and spatiotemporal consistency constraints. *IEEE Transactions on Cybernetics*, 2017. **48**(8): pp. 2485–2499.
19. Liu, Q., T. de Campos, W. Wang, P. Jackson, and A. Hilton. Person tracking using audio and depth cues. in *Proceedings of the IEEE International Conference on Computer Vision Workshops*. 2015.
20. Qian, X., A. Brutti, M. Omologo, and A. Cavallaro. 3D audio-visual speaker tracking with an adaptive particle filter. in *2017 IEEE International Conference on Acoustics, Speech and Signal Processing (ICASSP)*. 2017. IEEE.
21. Kumar, A., R. Jain, V. A. Devi, & A. Nayyar, (Eds.). *Object Tracking Technology: Trends, Challenges, Impact, and Applications*, 2023. Springer.
22. Gan, W., M.-S. Lee, C.-H. Wu, and C.-C.J. Kuo, Online object tracking via motion-guided convolutional neural network (MGNet). *Journal of Visual Communication and Image Representation*, 2018. **53**: pp. 180–191.

23. Zhu, S., Z. Fang, and F. Gao, Hierarchical convolutional features for end-to-end representation-based visual tracking. *Machine Vision and Applications*, 2018. **29**(6): pp. 955–963.
24. Shen, J., X. Tang, X. Dong, and L. Shao, Visual object tracking by hierarchical attention Siamese network. *IEEE Transactions on Cybernetics*, 2019. **50**(7): pp. 3068–3080.
25. Zhang, J., X. Jin, J. Sun, J. Wang, and A.K. Sangaiah, Spatial and semantic convolutional features for robust visual object tracking. *Multimedia Tools and Applications*, 2020. **79**(21): pp. 15095–15115.
26. Horn, B.K. and B.G. Schunck, Determining optical flow. *Artificial Intelligence*, 1981. **17**(1-3): pp. 185–203.
27. Dalal, N. and B. Triggs. Histograms of oriented gradients for human detection. in *2005 IEEE Computer Society Conference on Computer Vision and Pattern Recognition (CVPR'05)*. 2005. IEEE.
28. Bosch, A., A. Zisserman, and X. Munoz. Representing shape with a spatial pyramid kernel. in *Proceedings of the 6th ACM International Conference on Image and Video Retrieval*. 2007.
29. Song, S. and J. Xiao. Tracking revisited using RGBD camera: Unified benchmark and baselines. in *Proceedings of the IEEE International Conference on Computer Vision*. 2013.
30. Yan, S., J. Yang, J. Käpylä, F. Zheng, A. Leonardis, and J.-K. Kämäräinen. Depthtrack: Unveiling the power of rgbd tracking. in *Proceedings of the IEEE/CVF International Conference on Computer Vision*. 2021.
31. Gade, R. and T.B. Moeslund, Thermal cameras and applications: A survey. *Machine Vision and Applications*, 2014. **25**(1): pp. 245–262.
32. Luo, C., B. Sun, K. Yang, T. Lu, and W.-C. Yeh, Thermal infrared and visible sequences fusion tracking based on a hybrid tracking framework with adaptive weighting scheme. *Infrared Physics & Technology*, 2019. **99**: pp. 265–276.
33. Leykin, A., Y. Ran, and R. Hammoud. Thermal-visible video fusion for moving target tracking and pedestrian classification. in *2007 IEEE Conference on Computer Vision and Pattern Recognition*. 2007. IEEE.
34. Kim, H.-D., J.-S. Choi, and M.-S. Kim, Human-robot interaction in real environments by audio-visual integration. *International Journal of Control, Automation, and Systems*, 2007. **5**(1): pp. 61–69.
35. Zhang, D., Z. Zheng, M. Li, and R. Liu, CSART: Channel and spatial attention-guided residual learning for real-time object tracking. *Neurocomputing*, 2021. **436**: pp. 260–272.
36. Kim, H.-I. and R.-H. Park, Residual LSTM attention network for object tracking. *IEEE Signal Processing Letters*, 2018. **25**(7): pp. 1029–1033.
37. Leal-Taixé, L., C. Canton-Ferrer, and K. Schindler. Learning by tracking: Siamese CNN for robust target association. in *Proceedings of the IEEE Conference on Computer Vision and Pattern Recognition Workshops*. 2016.
38. Nousi, P., A. Tefas, and I. Pitas, Dense convolutional feature histograms for robust visual object tracking. *Image and Vision Computing*, 2020. **99**: p. 103933.
39. Gao, T., N. Wang, J. Cai, W. Lin, X. Yu, J. Qiu, and H. Gao, Explicitly exploiting hierarchical features in visual object tracking. *Neurocomputing*, 2020. **397**: pp. 203–211.

40. Thomas, M. and A.T. Joy, *Elements of Information Theory*. 2006. Wiley-Interscience.
41. Chen, Z., Y. Du, J. Deng, J. Zhuang, and P. Liu, Adaptive hyper-feature fusion for visual tracking. *IEEE Access*, 2020. **8**: pp. 68711–68724.
42. Elayaperumal, D. and Y.H. Joo, Robust visual object tracking using context-based spatial variation via multi-feature fusion. *Information Sciences*, 2021. **577**: pp. 467–482.
43. Yuan, D., X. Zhang, J. Liu, and D. Li, A multiple feature fused model for visual object tracking via correlation filters. *Multimedia Tools and Applications*, 2019. **78**(19): pp. 27271–27290.
44. Zhao, S., T. Xu, X.-J. Wu, and X.-F. Zhu, Adaptive feature fusion for visual object tracking. *Pattern Recognition*, 2021. **111**: p. 107679.

Chapter 4

Performance metrics for visual tracking

A qualitative and quantitative analysis

4.1 INTRODUCTION

Visual tracking has attracted many researchers and progressed significantly over the last decade. Tracking appearance models have evolved from hand-crafted features to features learned from deep neural networks. With the advancement in the field, every year many trackers are proposed and presented in various journals and conferences [1, 2]. To analyze the effectiveness of these trackers, there is a requirement for standard evaluation criteria. The standard evaluation metrics are essential not only to compare the performance of existing work, but also to progress further in the field.

Standardized potential performance metrics are significant to analyze and compare the performance of the tracker in real-time situations. The proposed new trackers are evaluating the tracking results on a limited set of evaluation metrics and compare them with similar algorithms. There is a deficiency in the similarity of evaluation metrics that makes it tough to analyze the tracking performance. There are plenty of evaluation metrics available, but selection of potential metrics remains a challenge [3]. Table 4.1 shows the recent work and the utilized sensors along with performance metrics utilized for performance comparison.

Experimental evaluation of the recent trackers is performed using both qualitative as well as quantitative analysis. Qualitative analysis analyzed the performance of the trackers for the various tracking attributes such as illumination variations, scale variations, deformations, fast motion, rotational, translation, out-of-view, and many more. These attributed performances analyzed the performance of the trackers by creating a bounding box (BB) on all the video frames of a sequence. The critical frames were identified in the complete video sequences and the performance of the trackers is evaluated by checking the percentage overlap between the tracker's BB and the ground truth BB. Tracker's drift is rigorously analyzed in the qualitative analysis to examine its performance in the presence of tedious environmental variations. On the other hand, quantitative evaluation is categorized either into performance evaluation using ground truth or without ground

DOI: 10.1201/9781003456322-4

Table 4.1 Representative work with a description of evaluation metrics used for comparing performance with state-of-the-art

SN	*References*	*Sensor*	*Algorithm*	*Evaluation metrics*	*Summary*
1.	Kumar et al. [4]	Vision	PF	CLE, F-Measure, AUC, OP, DP, FPS, Precision plot, and Success plot	Compared the tracking results against 13 other trackers.
2.	Lian [5]	Vision	CF	CLE, OR, FPS, and Success rate plot	Results compared against 4 other state-of-the arts.
3.	Taylor et al. [6]	Vision	Kalman filter	MDE, Success rate, and APT	No performance comparison.
4.	Walia et al. [7]	Vision	PF	CLE, F-Measure, Precision plot, and Success plot	Results compared against 9 other state-of-the-arts.
5.	Wu et al. [8]	Vision	Incremental PCA	CLE, AUC, Precision plot, and Success plot	Results compared against 10 other state-of-the-arts.
6.	Xu et al. [9]	Vision	Discriminative CF	CLE, AUC, OP, DP, FPS, EAO, Accuracy, Robustness, Stability analysis, Precision plot, and Success plot	Results compared against 8 other state-of-the-arts.
7.	Zhang et al. [10]	Vision	Siamese networks	AUC, Precision, OP, DP, FPS, EAO, Accuracy, Robustness, Tracking speed, Precision plot, and Success plot	Results compared against 12 other state-of-the-arts.
8.	Zhao et al. [11]	Vision	CNN	AUC, Precision, Normalized precision, Success, Tracking speed, EAO, Accuracy, Robustness, Precision plot, and Success plot	Results compared against 8 other state-of-the-arts.

9.	Zheng et al. [12]	Vision	Deep dual CF	AUC, Precision, Normalized precision, Success rate, FPS, Speed, EAO, Accuracy, Robustness, TPR, TNR, Precision plot, and Success plot	Results compared against 8 other state-of-the-arts.
10.	Xiao et al. [13]	Vision & Depth	Part-based tracking	AUC, Failure, Accuracy	Results compared against 4 other state-of-the-arts.
11.	Ma et al. [14]	Vision & Depth	Sparse learning	CLE, Success rate, ACPE	Results compared against 4 other state-of-the-arts.
12.	Luo et al. [15]	Vision & TIR	CF	EAO, Accuracy, Robustness, MPR and MSR plot	Results compared against 10 other state-of-the-arts.
13.	Liu et al. [16]	Vision & TIR	Siamese networks	FPS, EAO, Accuracy, Robustness, Precision plot, and Success plot	Results compared against 18 other state-of-the-arts.

Note: TIR: Thermal Infrared, PF: Particle filter, CF: Correlation filter, CNN: Convolutional neural network, CLE: Center location error, AUC: Area under the curve, OP: Overlap precision, DP: Distance precision, FPS: Frames/sec, EAO: Expected accuracy overlap, MDR: Mean distance error, APE: Average processing time, ACPE: Average center position error, MPR: Maximum precision rate, MSR: Maximum success rate.

truth. The next section will detail these categories of tracker's evaluation in detail.

4.2 PERFORMANCE METRICS FOR TRACKER EVALUATION

Performance metrics focus to evaluate the tracker in critical scenarios to estimate its robustness and reliability in the real-world. The tracking performance measures aim to identify the tracker's failure cases too so that the gaps can be filed before its practical implementation. The quantitative performance measures are either considering ground truth data or ignoring it in case of its non-availability. Manual annotations techniques have been proposed to find the centroid using BB on the ROI or the target [3]. Figure 4.1 presents the centroid C_t (x_t, y_t) computation from the target region $\mathcal{R}_t$ for ground truth BB (BB^G) at time t. The next section will elaborate on the various categories of performance metrics.

Figure 4.2 illustrates the overlap between the tracker's BB (BB^T) and ground truth BB (BB^G) for the jogging1 video sequence of OTB data sets [17]. Tracker's true positive (TP), true negative (TN), false positive (FP) and false negative (FN) are represented as the overlap ratio between the ground truth BB and tracker's BB.

4.3 PERFORMANCE METRICS WITHOUT GROUND TRUTH

Earlier, due to the non-availability of ground truth data tracking performance has been evaluated by using the posterior density considering the time-reversible nature of object trajectory [18], irregularities in the pattern of the target's appearance [19] and the tracker's uncertainty to recover from failures [20].

In this direction, Wu et al. [18] have proposed an evaluation criterion based on the comparison of prior and posterior density probability distribution exhibited by the target's motion. Prior is used to initialize the tracker

Figure 4.1 Representing the centroid C_t on the target region $\mathcal{R}_t$ at time t for ground truth BB (BB^G) on sample video frames from OTB data sets [17].

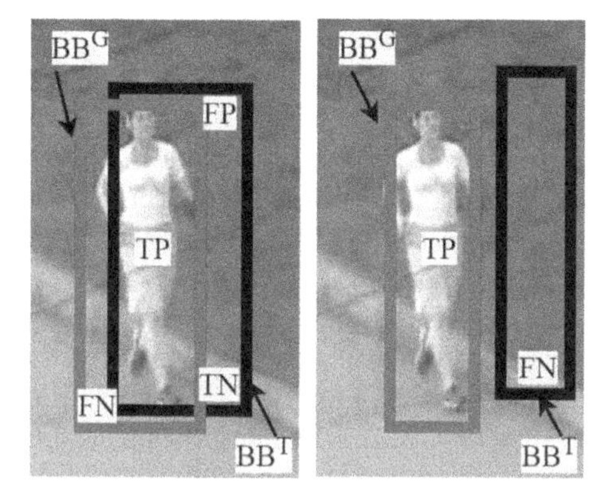

Figure 4.2 Video frames from *jogging1* sequence of OTB data set representing the overlap between the tracker's BB (BB^T) and ground truth BB (BB^G).

while posterior is used in the time-reversed chain to evaluate the tracker's statistics. This methodology is generic and not specific to any tracking algorithm. However, the evaluation results are not reliable for long-term videos. To address this, authors have proposed to determine the tracker's recovery rate from failures considering the time-reversibility constraint [20]. Tracker stability is determined by the successful tracking results, whereas the time-reversibility constraint measures the tracker's uncertainty from tracking failures using a reverse-tracker. The similarity between forward and reverse tracking is used to compute the tracking quality. In another direction, authors have proposed a probabilistic approach to evaluate tracking algorithms without ground truth [19]. Features, namely shape, appearance, and motion, are aggregated in a naïve Bayesian classifier to compute probability scores. This probability score is used to evaluate the performance of the trackers. However, Chau et al. [21] proposed a set of seven features to determine the motion trajectory of the moving target. Correct and erroneous tracking trajectories are distinguished by providing a local score between zero and one for each considered feature. Finally, a global score is computed by integrating the local score from all the features. This global score varies between zero and one and is used to determine the tracking quality.

Furthermore, Wu and Zheng [22] have proposed automatic performance metric criteria considering the target's shape and appearance similarity along with the tracker's uncertainty without using ground truth information. Five independent ambiguity tests have been conducted and a weighted fused score is obtained to compute comprehensive tracking results. Each test is evaluated for uncertainty and a comprehensive score is checked to re-initialize the tracker if it falls below a predicted threshold. However, the authors have proposed metrics based on the difference in color and motion along the boundary of the estimated target [23]. Color and motion are measured for temporal localization whereas their intra and inter-frame scores from the boundary are used for spatial localization. In ref. [24], the authors have utilized several low-level features along with a feedback mechanism to track the sub-contours of the target in each frame. The boundary estimation is improved by incorporating the energy terms in the estimation process. The weights of energy terms are updated with the performance metrics to improve accuracy. Pang and Ling [25] have evaluated the tracker's performance from a different perspective. The authors have collected the previously published algorithms and ranked them using four different methodologies. They aim to identify the incorporated subjective biases in the results of the published work. However, the analysis lacks a comparison of the performance metrics utilized in the various studies. In sum, performance metrics without ground truth can compute the performance of the tracker based on certain other parameters but the results are not sufficiently reliable to predict the real-time tracking results in presence of complex environmental variations.

4.4 PERFORMANCE METRICS WITH GROUND TRUTH

Initially, performance metrics with ground truth are proposed in PETS (Performance Evaluation of Tracking and Surveillance) workshop. The workshop aims to give effective performance metrics to compare tracking results independent of tracking algorithms. Details for performance metrics considered ground truth information are followed in turn.

4.4.1 Center location error (CLE)

CLE is one of the oldest, simplest, and most accurate performance metrics. Due to this, it has been exploited by many published works [4, 5, 7, 8]. CLE (δ_t) computes the Euclidean distance between the centroid predicted by the tracker's BB $\left(C_t^T\right)$ and the ground truth BB $\left(C_t^G\right)$ using Eq. (4.1).

$$\delta_t = \sqrt{\frac{1}{N}\sum_{t=1}^{N}\left\|C_t^G - C_t^T\right\|} \tag{4.1}$$

where N is the total number of frames in the video sequence. This performance metric is solely based on the target's centroid computation. Target's centroid varies when the target scale changes abruptly and leads to performance failures. To address this, target size is incorporated into CLE to propose normalized CLE [26]. The normalized CLE $\left(\delta_t^n\right)$ is computed using Eq. (4.2).

$$\delta_t^{\eta} = \sum_{t=1}^{N} \left\| \frac{C_t^G - C_t^T}{S\left(\mathcal{R}_t^G\right)} \right\| \tag{4.2}$$

where $S\left(\mathcal{R}_t^G\right)$ is the target's size predicted by the tracker. However, this may compute false results as CLE may reduce according to the target size.

4.4.2 F-measure

F-measure is an evaluation metric used to measure the effectiveness of the tracker's performance [27]. It is a simple object-based assessment metric considering the overlap between the tracker's BB and the ground truth BB. F-Measure is recently used by many researchers to compare tracker's performance [4, 7]. It is determined as a harmonic mean between the precision (P_t) and recall (R_t) values at time t using Eq. (4.3).

$$\mathcal{F}_t = \frac{2 \times P_t \times P_t}{P_t + R_t} \tag{4.3}$$

where P_t is the overlap of the tracker's BB (BB^T) with ground truth BB (BB^G) w.r.t. to tracker BB (BB^T) computed using Eq. (4.4) and R_t is the overlap of the tracker's BB (BB^T) with ground truth BB (BB^G) w.r.t. to ground truth BB (BB^G) computed using Eq. (4.5).

$$P_t = \left| \frac{BB^T \cap BB^G}{BB^T} \right| \tag{4.4}$$

$$R_t = \left| \frac{BB^T \cap BB^G}{BB^G} \right| \tag{4.5}$$

4.4.3 Distance precision, overlap precision, and area under the curve

Distance precision is computed as the corresponding value to the precision thresholds at a certain pixel level. Overlap precision is the measure of the frame percentage where the BB overlap is above a pre-defined threshold. Area under the curve (AUC) is the measure of the percentage of the successfully tracked frames evaluated over various overlap thresholds.

4.4.4 Expected accuracy overlap, robustness, and accuracy

Expected average overlap (EAO) defines the accuracy and failures in a frame in an integrated manner. It computes the overlap between accuracy and failure in a video sequence.

Robustness computes the number of times the tracker lost the target. It defines the average failure rate obtained by averaging the results of different runs of the tracker. When the overlap between the tracked BB and the ground truth BB becomes zero, the target is considered to be lost by the tracker.

Accuracy measures the performance quality of the tracker. It keeps track of the successful tracking results in the entire video. It is defined as the overlap between the tracked BB and the ground truth BB in the entire video sequence.

4.4.5 Performance plots

The performance of the tracker is also plotted using precision plots and success plots. Precision plot represents the mean precision in accordance with various CLE thresholds. In other words, it measures the number of frames having a distance between predicted results and ground truth below a certain number of pixels.

On the other hand, the success plot defines the percentage of successfully tracked frames on various overlap thresholds (in %). The predicted frame is considered to be successfully tracked if the overlap between the predicted frame and the ground truth varies over a pre-defined threshold.

4.5 SUMMARY

In this chapter, we have discussed the various performance metrics used for evaluating the performance of single object trackers. Evaluation metrics are essential for the fair evaluation of the tracker's accuracy and robustness. Generally, the ground truth data is provided with the data set to compare the predicted tracking results with the actual tracking outcomes. However, the motion patterns and target appearance variations are analyzed during the non-availability of the ground truth data.

Evaluation metrics provide a common base to compare the performance of various modern equipment. Robustness, accuracy, and failure evaluation are performed to check the practical applicability of the trackers.

REFERENCES

1. Kumar, A., G.S. Walia, and K. Sharma, Recent trends in multicue based visual tracking: A review. *Expert Systems with Applications*, 2020. **162**: p. 113711.

2. Kumar, A., R. Jain, V. A. Devi, & A. Nayyar, (Eds.). *Object Tracking Technology: Trends, Challenges, Impact, and Applications*, 2023. Springer.
3. Čehovin, L., A. Leonardis, and M. Kristan, Visual object tracking performance measures revisited. *IEEE Transactions on Image Processing*, 2016. **25**(3): pp. 1261–1274.
4. Kumar, A., G.S. Walia, and K. Sharma, Real-time visual tracking via multi-cue based adaptive particle filter framework. *Multimedia Tools and Applications*, 2020. **79**(29): pp. 20639–20663.
5. Lian, G.-Y., A novel real-time object tracking based on kernelized correlation filter with self-adaptive scale computation in combination with color attribution. *Journal of Ambient Intelligence and Humanized Computing*, 2020: pp. 1–9.
6. Taylor, L.E., M. Mirdanies, and R.P. Saputra, Optimized object tracking technique using Kalman filter. arXiv preprint arXiv:2103.05467, 2021.
7. Walia, G.S., A. Kumar, A. Saxena, K. Sharma, and K. Singh, Robust object tracking with crow search optimized multi-cue particle filter. *Pattern Analysis and Applications*, 2020. **23**(3): pp. 1439–1455.
8. Wu, F., C.M. Vong, and Q. Liu, Tracking objects with partial occlusion by background alignment. *Neurocomputing*, 2020. **402**: pp. 1–13.
9. Xu, T., Z. Feng, X.-J. Wu, and J. Kittler, Adaptive channel selection for robust visual object tracking with discriminative correlation filters. *International Journal of Computer Vision*, 2021. **129**(5): pp. 1359–1375.
10. Zhang, D., Z. Zheng, M. Li, and R. Liu, CSART: Channel and spatial attention-guided residual learning for real-time object tracking. *Neurocomputing*, 2021. **436**: pp. 260–272.
11. Zhao, S., T. Xu, X.-J. Wu, and X.-F. Zhu, Adaptive feature fusion for visual object tracking. *Pattern Recognition*, 2021. **111**: p. 107679.
12. Zheng, Y., X. Liu, X. Cheng, K. Zhang, Y. Wu, and S. Chen, Multi-task deep dual correlation filters for visual tracking. *IEEE Transactions on Image Processing*, 2020. **29**: pp. 9614–9626.
13. Xiao, J., R. Stolkin, Y. Gao, and A. Leonardis, Robust fusion of color and depth data for RGB-D target tracking using adaptive range-invariant depth models and spatio-temporal consistency constraints. *IEEE Transactions on Cybernetics*, 2017. **48**(8): pp. 2485–2499.
14. Ma, Z.-A. and Z.-Y. Xiang, Robust object tracking with RGBD-based sparse learning. *Frontiers of Information Technology & Electronic Engineering*, 2017. **18**(7): pp. 989–1001.
15. Luo, C., B. Sun, K. Yang, T. Lu, and W.-C. Yeh, Thermal infrared and visible sequences fusion tracking based on a hybrid tracking framework with adaptive weighting scheme. *Infrared Physics & Technology*, 2019. **99**: pp. 265–276.
16. Liu, Q., X. Li, Z. He, N. Fan, D. Yuan, and H. Wang, Learning deep multilevel similarity for thermal infrared object tracking. *IEEE Transactions on Multimedia*, 2021. **23**: pp. 2114–2126.
17. Wu, Y., J. Lim, and M.-H. Yang. Online object tracking: A benchmark. in *Proceedings of the IEEE Conference on Computer Vision and Pattern Recognition*. 2013.
18. Wu, H., A.C. Sankaranarayanan, and R. Chellappa, Online empirical evaluation of tracking algorithms. *IEEE Transactions on Pattern Analysis and Machine Intelligence*, 2009. **32**(8): pp. 1443–1458.

19. Spampinato, C., S. Palazzo, and D. Giordano. Evaluation of tracking algorithm performance without ground-truth data. in *2012 19th IEEE International Conference on Image Processing*. 2012. IEEE.
20. SanMiguel, J.C., A. Cavallaro, and J.M. Martínez, Adaptive online performance evaluation of video trackers. *IEEE Transactions on Image Processing*, 2012. **21**(5): pp. 2812–2823.
21. Chau, D.P., F. Bremond, and M. Thonnat. Online evaluation of tracking algorithm performance. in *3rd International Conference on Imaging for Crime Detection and Prevention (ICDP 2009)*. 2009. IET.
22. Wu, H. and Q. Zheng, Self-evaluation for video tracking systems. 2004. Maryland Univ. College Park Dept. of Electrical and Computer Engineering.
23. Erdem, C.E., A.M. Tekalp, and B. Sankur. Metrics for performance evaluation of video object segmentation and tracking without ground-truth. in *Proceedings 2001 International Conference on Image Processing (Cat. No. 01CH37205)*. 2001. IEEE.
24. Erdem, Ç.E., B. Sankur, and A.M. Tekalp. Non-rigid object tracking using performance evaluation measures as feedback. in *Proceedings of the 2001 IEEE Computer Society Conference on Computer Vision and Pattern Recognition. CVPR 2001*. 2001. IEEE.
25. Pang, Y. and H. Ling. Finding the best from the second bests-inhibiting subjective bias in evaluation of visual tracking algorithms. in *Proceedings of the IEEE International Conference on Computer Vision*. 2013.
26. Smeulders, A.W., D.M. Chu, R. Cucchiara, S. Calderara, A. Dehghan, and M. Shah, Visual tracking: An experimental survey. *IEEE Transactions on Pattern Analysis and Machine Intelligence*, 2013. **36**(7): pp. 1442–1468.
27. Lazarevic-McManus, N., J. Renno, D. Makris, and G.A. Jones, An object-based comparative methodology for motion detection based on the F-Measure. *Computer Vision and Image Understanding*, 2008. **111**(1): pp. 74–85.

Chapter 5

Visual tracking data sets

Benchmark for evaluation

5.1 INTRODUCTION

In order to evaluate the tracking performance on a standardized set, there is a requirement for data sets that can evaluate the tracking performance in presence of tedious environmental variations [1, 2]. With the advancement in object tracking, more than ten trackers are proposed every year. The strength of these algorithms needs to be compared to complex videos to analyze their behavior in real time. Visual tracking data sets are categorized into self-generated data sets [3] and public data sets [4–6]. Each of these data sets has its benefits with respect to visual tracking problems.

Earlier, the performance of tracking algorithms was evaluated either on fewer sequences or on a few frames from a video sequence [7]. The results are not sufficiently reliable to analyze the tracker's performance and have shown tracking failures during practical applications. For a comprehensive evaluation, the tracking algorithms must be executed not only on entire data sets but also on multiple data sets [8]. This performance evaluation is crucial to identify the tracker's drift in tedious environmental conditions.

Tracking performance is severely impacted by various environmental attributed challenges that include illumination variations, scale variations, fast motion, motion blur, occlusion, background clutters, and target rotations. There are many public data sets have been proposed attributing these challenges to their video data sets [2, 9]. These data sets are primarily for visual trackers based on vision sensors. Typical tracking challenges such as rain, fog, and day and night conditions are not considered in the video sequences of these data sets. To enhance tracker capability, visual data can be integrated with data from specialized sensors such as thermal, IR, and depth for robust tracking solutions. It has been well analyzed that the tracking performance can be boosted for day and night by integrating thermal data. Public multi-modal data sets have been proposed to compare the performance of RGB-T trackers [10, 11]. The limitations of these multi-modal data sets are: (1) significantly fewer videos; (2) insufficient duration; and (3) fewer attributed challenges are considered. To address this, large diversified multi-modal data sets have been proposed [12, 13]. These data sets consist

DOI: 10.1201/9781003456322-5

of a large number of video frames and precise alignment between the RGB and thermal pairs. Apart from thermal data, depth information is also integrated into the trackers to handle occlusion. The performance of these trackers is compared using RGBD data sets [14, 15]. The RGB and depth image pairs are synchronized to avoid uncertainty in the ground truth information. The videos are recorded either with a stationary or moving camera. The length of collected videos is sufficient to analyze the tracker's performance under tedious tracking challenges.

To address the need for data-hungry DL-based trackers, large data sets have been proposed [16, 17]. These data sets contain sufficient long-duration videos with multiple attributed challenges. High-quality dense rich annotations, manual labeling, and a wide range of diversified challenges are some generic features of these data sets. These data sets include a lot of video frames having out-of-view and small-sized targets. Also, multi-modal large data sets are proposed to compare the trackers utilizing multi-modal features [13, 18]. These data sets not only include challenging videos but also consider weather situations. In [18], data sets have videos captured in summer as well as in the winter season. Real-time complex scenarios are integrated for the exhaustive analysis of the trackers. In summary, large data sets with a wide range and long videos are sufficient to provide enough data for the training and testing of deep learning trackers. These data sets include tedious real-time complex tracking variations suitable to test the trackers before their practical implication.

5.2 PROBLEMS WITH THE SELF-GENERATED DATA SETS

Some researchers have proposed and analyzed the performance of their trackers on self-generated data sets [3]. Self-generated data sets are created by the researchers considering various challenging attributes to test the tracking algorithms. The details of these videos are not made public by the developer. The videos are used solely for the analysis of tracking algorithms designed by the researchers themselves. However, there is no means to identify the correctness of the proposed data sets as the annotations are not publicly available. It has been argued that self-generated data sets lack fair comparison as they are tested for a single algorithm only.

5.3 SALIENT FEATURES OF VISUAL TRACKING PUBLIC DATA SETS

For efficient tracking solutions, public data sets have been proposed. For simplicity and clarity, we have categorized the single object tracking data sets into traditional tracking data sets and multi-modal tracking data sets. The salient features of the traditional tracking data set are as follows in turn.

5.3.1 Data sets for short-term traditional tracking

The short-term data sets for traditional tracking are quite popular among researchers [19]. These data sets contain a sufficient number of videos with a shorter duration which is enough to calibrate the tracking algorithms. These data sets are widely used to compare and analyze the tracking performance in tough environmental situations. Table 5.1 shows the salient features of recent benchmarks which can be exploited for traditional tracking. The details of data sets under this category are as follows.

- VIRAT [20] contains 17 videos with a total duration of 29 hours at 24 FPS. The data set contains diversified videos with 23 different event types. It contains both stationary grounds as well as aerial videos. The stationary ground videos consist of 25 hours in total with an average of 1.6 hours of videos per scene. Whereas a total of 25 hours of aerial videos are recorded, of which 4 hours of videos are selected manually and added to the data set. Objects are annotated with rectangular BB marking objects in the scene.
- ALOV300++ [6] contains 315 short videos under diverse tracking circumstances at 30 FPS. The average duration of videos varies from 9.2 seconds to 35 seconds. A collection of 10 is also included having a duration between 1 to 2 minutes. Most videos are real-time and collected from YouTube with 64 different objects. The ground truth is obtained by interpolation methods for the intermediate frames.
- TColor128 [9] consists of 128 RGB sequences with varying degrees of attributed challenges at 30 FPS. Videos captured by the stationary camera are collected from the Internet and previous works. Videos collected from the Internet are rich in eleven complex tracking challenges. Ground truth information is provided in text for evaluation.
- OTB-2015 [2] is the most popular and widely used public benchmark for single object tracking. It contains 100 real-time videos captured by the stationary camera at 30 FPS. Most videos have humans as the target with 36 bodies and 26 face/heads. If one video has two targets, then those are considered two different videos. A single video has more than one attributed challenge to make it tedious and equivalent to real-time situations. The annotated video ground truth data is provided in a text file.
- UAV123 [4] has 123 fully annotated new HD videos captured by a moving camera mounted on UAV (unmanned aerial vehicle) at a frame rate between 30 and 90 FPS. All video sequences are 720 px captured from low altitudes varying from 5–25 meters. Videos are captured from a variety of scenes such as urban, buildings, roads, beaches, and many more with a wide range of targets. In addition, synthetic videos in which the target move along pre-determined trajectories with different backgrounds rendered by a game engine are also included to add diversity to the data set.

Table 5.1 Benchmark for performance evaluation for short-term traditional tracking

SN	References	Data set	Year	Web-link	Sequences	Frames	Attributed challenge	Summary
1.	Oh et al. [20]	VIRAT	2011	http://www.viratdata.org/	17	1500	23	Complex video sequences with diverse multiple aspects.
2.	Smeulders et al. [6]	ALOV300++	2013	http://alov300pp.joomlafree.it/	314	151k+	10+	Real-life YouTube videos with 64 different categories.
3.	Liang et al. [9]	TColor128	2015	https://www3.cs.stonybrook.edu/~hling/data/TColor-128/TColor-128.html#dataset	128	55k+	11	Collection of videos from the Internet and previous data sets with real-time situations.
4.	Wu et al. [2]	OTB 2015	2015	http://cvlab.hanyang.ac.kr/tracker_benchmark/datasets.html	100	59k+	11	Videos with fully annotated ground truth with challenging aspects.
5.	Mueller et al. [4]	UAV123	2016	https://cemse.kaust.edu.sa/ivul/uav123	123	112k+	12	Video captured with low altitude UAVs with emphasis on scale variations.
6.	Kiani Galoogahi et al. [21]	NfS	2017	http://ci2cv.net/nfs/index.html	100	380k+	9	Videos captured at higher FPS in real-world scenarios.
7.	Li and Yeung [22]	DTB70	2017	https://github.com/flyers/drone-tracking	70	18k+	11	Contains videos recorded by drones and taken from YouTube.
8.	Kristan et al. [23]	VOT 2018	2018	https://www.votchallenge.net/vot2018/dataset.html	60	14k+	9	Videos collected from real-life situations.
9.	Du et al. [24]	UAVDT	2018	https://sites.google.com/site/daviddo0323/	100	80k	14	Videos captured with UAV at multiple locations with tedious tracking challenges.
10.	Yu et al. [25]	UAV benchmark	2020	https://sites.google.com/site/daviddo0323/	100	80k	16	Videos recorded in a real-time complex environment.

Note: Data sets are arranged in ascending order of the published year.

- NfS (Need for Speed) [21] consists of 100 videos at a higher frame rate of 240 FPS. Out of 100 videos, 75 are captured by iPhone 6 (or above) or iPad Pro and the remaining are gathered from YouTube. More than 17 types of targets are included in the videos for tracking. The data set is suitable to analyze the performance of the fast trackers.
- DTB70 [22] consists of 70 videos of 1280 × 720 resolution captured by a drone camera. Videos are either captured by motion-controlled drone camera or taken from YouTube. Three categories of target objects are considered, namely humans, animals, and rigid. The videos are rich in a wide range of tracking challenges with manually annotated ground truth.
- VOT 2018 [23] has in total of 70 videos at frame rates varying from 25 to 30 FPS. Out of 70 videos, 35 videos are taken from LTB35 data sets [26], 20 from UAVL20 [4], 3 from [27], 6 from YouTube, and 6 are generated by AMP [28]. The data set is rich in the out-of-view challenge in a total of 463 target disappearances. The video's resolution varies between 1280 × 720 and 290 × 217. Targets are annotated by axis-aligned bounding boxes.
- UAVDT [24] contains 100 fully annotated videos of 10 hours duration at 30 FPS. Videos are captured from multiple locations, such as urban, toll, crossing, T-point, and many more with a resolution of 1080 × 540. Videos are recorded by UAV from varying altitudes between 10 to above 70 categorized as low, medium, and high.
- UAV benchmark [25] consists of 100 videos at 30 FPS having 2.7k vehicles. Videos are recorded in complex real-time scenarios with manually annotated bounding boxes. Mostly sequences comprise of 4–5 attributed challenges to make tracking tedious.

5.3.2 Multi-modal data sets for multi-modal tracking

In recent years, visual tracking using vision sensors has shown extreme progress. The reason is easy to feature extraction and simple computation. However, trackers based on visual features have shown limited performance during the night and heavy occlusion scenarios. To further extend the capabilities of the vision trackers, features extracted from specialized sensors are integrated with RGB data [29, 30]. To evaluate the performance of these multi-modal trackers, there is a need for multi-modal tracking data sets which can analyze, evaluate, and compare the performance of these trackers. The salient features of multi-modal tracking data sets are tabulated in Table 5.2 and details are as follows.

- AV16.3 data set [31] consists of 8 annotated audio and video sequences recorded in a synchronized manner having an average duration of 30–511 seconds. In total, the data set has three videos DIVX AVI files

Table 5.2 Benchmark for performance evaluation for multi-modal trackers

SN	*Reference*	*Data set*	*Year*	*Web link*	*Sequences*	*Frames*	*Attributed challenges*	*Summary*
1.	Lathoud et al. [31]	AV 16.3	2004	https://www.idiap.ch/dataset/av16-3	42	200	—	16 microphones along with 3 cameras to record video and voice data set.
2.	Song and Xiao [32]	PTB	2013	http://tracking.cs.princeton.edu/dataset.html	100	55k+	11	Consisting of both RGB and depth video data sets manually annotated.
3.	Wu et al. [33]	TIV	2014	http://csr.bu.edu/BU-TIV/BUTIV.html	16	63k+	7	7 different scenes along with 2 indoor videos.
4.	Hwang et al. [34]	KAIST	2015	http://rcv.kaist.ac.kr/multispectralpedestrian/	—	95k+	4	Captured video during day and night using a camera mounted on the car.
5.	Felsberg et al. [35]	VOT-TIR2016	2016	http://www.votchallenge.net/vot2016/dataset.html	25	17k+	14	Video was captured from 10 different sensors from 9 different sources.
6.	Patino et al. [11]	PETS 2016	2016	http://www.cvg.reading.ac.uk/PETS2016/a.html	11	—	Mainly three	Multi-sensor data sets having RGB and thermal sequences.
7.	Gonzales et al. [36]	CVC-14	2016	http://adas.cvc.uab.es/elektra/enigmaportfolio/cvc-14-visible-fir-daynight/pedestrian-sequence-dataset/	2	48k+	—	The data set consists of video from RGB, IR, and RGB plus IR.
8.	Palmero et al. [37]	VAP Trimodal	2016	http://www.vap.aau.dk/dataset/	3	11k+	—	Indoor videos with pixel-level manual annotation.
9.	Li et al. [10]	GTOT	2016	http://chenglongli.cn/people/lcl/datasetcode.html	100	15k+	7	RGB-Thermal video pairs with ground truth annotations.
10.	Moyà-Alcover et al. [14]	GSM	2017	http://gsm.uib.es/dataset	7	3.3k+	—	Both RGB and depth data sets with handheld annotation.
11.	Xiao et al. [15]	BTB	2018	https://research.birmingham.ac.uk/portal/en/datasets/search.html	36	20k+	10	Each video frame has two types of annotation namely, binary, and statistical.

Note: Data sets are arranged in ascending order of the published year.

and sixteen audio WAV files sampled at 25 Hz and 16 kHz, respectively. Three cameras are used to capture the 3-D locations and two arrays of 8-microphones separated by a distance of 0.8m are used to record the audio files.

- PTB (Princeton tracking benchmark) data set [32] is one of the most popular multi-modal data sets having sequences with RGB and depth data captured by a stationary camera. Depth data is captured by the Microsoft Kinect depth sensor. Large numbers of videos with multiple attributed challenges are captured to ensure the regressive performance analysis. However, most of the videos are taken in similar background situations with the same target which makes the data set less diversified. All frames are manually annotated.
- TIV (Thermal infrared video) benchmark [33] contains 16 videos from seven different indoor and outdoor scenes at the high resolution of 1024 × 1024. Each pixel in a video frame has a temperature range varying from 3000 to 7000 units. Sequences have a frame rate between 5 and 131 FPS depending on the target's speed.
- KAIST [34] is a multispectral data set having well-aligned RGB-thermal image pairs. RGB data is captured by PointGrey Flea3 shuttered color camera and thermal data by FLIR-A35 thermal camera using a beam splitter. RGB and thermal data have resolutions of 640 × 480 and 320 × 256, respectively, and a frame rate of 20 FPS. Data frames are manually annotated using a computer vision toolbox for better alignment between the RGB and thermal data.
- VOT-TIR 2016 [35] has thermal-infrared sequences captured from nine different sources using ten different sensors. Videos are either captured outdoors in real-life scenarios or indoors creating a virtual challenging environment. The average duration of the sequence is 240 frames with resolution varying between 305 × 225 and 1920 × 480 pixels. Data frames are annotated for global and local attributed challenges.
- PETS 2016 [11] contains two data sets, namely multi-sensor ARENA data set and new IPATCH data set. ARENA data sets have 22 video recordings from multi-camera of 1280 × 960 resolutions at 30 FPS. IPATCH data set has fourteen videos recorded by multi-sensors, i.e. visible and thermal. RGB data is recorded by four AXIS P1427-E Network vision sensors at a 30 FPS frame rate. Three thermal cameras, two FLIR SC655 and one FLIR A65 have recorded videos at 25 FPS and 30 FPS, respectively. For thermal cameras, environmental temperature varies between –15 to +50 degree Celsius.
- CVC-14 [36] contains two long sequence pairs recorded by the vision sensor and IR sensor. One video is recorded during the day and the other during the night. IDS UI-3240CP camera has a resolution of 1280 × 1024 with having 60 FPS frame rate and FLIR Tau 2 has a 640 × 512 resolution at 30/25 Hz frame rate.

- VAP Trimodal [37] contains three indoor video sequences captured by vision, thermal, and depth of 640 × 480 resolution at 15 FPS. RGB and depth data are recorded by Microsoft Kinect XBOX360 and thermal data by AXIS Q1922. Multi-sensor data is calibrated using a camera calibration toolbox to align the vision, depth, and thermal information. The data set is annotated using a pixel annotator.
- GTOT [10] has 50 video pairs for grayscale and thermal data under different real-time conditions. The self-developed annotation tool is used to align the grayscale and thermal images manually. All video pairs are manually annotated for ground truth by a single person only.
- GSM (Generic scene modeling) [14] contains seven sequences having RGB and depth data. Each sequence has a specific test objective in which the first 100 frames are used for training purposes. The average length of the sequence varies from 200 to 1231 frames. Out of the total 3361 frames, only 87 frames are with ground truth annotated manually.
- BTB (University of Birmingham RGB-D data set) [15] consists of 36 video sequences with average lengths varying between 300 and 700 frames. Two Asus Xtion RGB-D sensors are used to record the videos. One sensor is stationary while the other is moving along the trajectory. The recorded average target's depth is between 0.5 and 8 meters. The frames are manually annotated for ground truth.

5.4 LARGE DATA SETS FOR LONG-TERM TRACKING

An efficient tracker can localize the target for a sufficient duration in which the target may be occluded, out-of-view and/or re-enter the scene multiple times. However, most of the earlier data sets have primarily focused on one attributed challenge per video sequence [6]. If multiple attributed challenges are captured in single videos, the duration of the video is not long enough [2]. Also, the number of videos in the data sets is not sufficient to cater for real-time tracking performance. In addition, DL-based trackers require a lot of training data emphasizing the need for large data sets. Hence, data sets with longer video sequences are required not only for real-time testing of the trackers but also to satisfy the data requirements of DL-based trackers. Table 5.3 shows the salient parameters of recently published large data sets that can be used for long-term tracking. The in-depth details of these benchmarks are as follows.

- InfAR (Infrared action recognition) data set [18] is a large data set with 600 videos captured by the vision and IR sensor having 12 common human actions. The average length of the video is four seconds of 293 × 256 resolutions at 25 FPS. Each video has a single person

Table 5.3 Large data sets for performance evaluation for deep learning-based trackers

SN	*Reference*	*Data set*	*Year*	*Web-link*	*Sequences*	*Frames*	*Attributed challenges*	*Summary*
1.	Gao et al. [18]	InfAR	2016	https://sites.google.com/site/gaochenqiang/publication/infraredaction-dataset	600	600k	12	12 action classes with 50 video clips in each class.
2.	Li et al. [13]	RGBT210	2017	http://chenglongli.cn/code-dataset/	420	210k	12	Videos were captured using a moving camera along with occlusion annotation.
3.	Real et al. [38]	YT-BB	2017	https://research.google.com/youtube-bb/	380k	5.6M	–	23 object classes with human-annotated BB.
4.	Muller et al. [5]	TrackingNet	2018	https://github.com/SilvioGiancola/TrackingNet-devkit	30k	14M+	15	27 object classes with train and test split video sequences.
5.	Huang et al. [1]	GOT-10k	2019	http://got-10k.aitestunion.com/downloads	10k	1.5M	6	Large and diversified data set with 563 object classes.
6.	Li et al. [12]	RGBT234	2019	https://sites.google.com/view/ahutracking001/	234	234k	12	Large-scale data set with alignment between RGB and thermal pair with occlusion annotation.
7.	Zhu et al. [17]	VisDrone2019	2019	https://github.com/VisDrone/VisDrone-Dataset	263	179k+	4+	Wide range of aspects in a large geographical location is covered in a video using a drone.
8.	Lukežič et al. [39]	CBDT	2019	https://paperswithcode.com/dataset/cdtb	80	101k	13	Largest and diversified data set with RGB and depth data.
9.	Fan et al. [16]	LaSOT	2021	https://cis.temple.edu/lasot/	1400	3.52M	14	Large-scale and high-quality videos with 70 object classes.
10.	Liu et al. [40]	LSOTB-TIR	2023	https://github.com/QiaoLiuHit/LSOTB-TIR	1416	643k	12	The data set is split into two parts to address short-term tracking as well as long-term tracking.
11.	Feng et al. [41]	DTTD	2023	https://github.com/augcog/DTTDv1	103	55k	Multiple	3D data set having real and synthetic images along with train test split.

Note: Data sets are arranged in ascending order of the published year.

or multiple persons performing similar or different actions. The most video focus on occlusion challenges along with complex background and viewpoint variations.

- RGBT210 [13] consists of a large number of RGB and thermal video pairs captured by both static and moving cameras. Each video is sufficiently long with an average length of 8k frames per video. Vision data is recorded by SONY ExView HAD CCD camera and thermal information is captured by DLS-H37DM-A thermal IR imager. All videos are annotated manually by a single person. All occlusion levels, i.e. no, full, and partial are annotated for fair evaluation during the challenge in the video sequence.
- YT-BB (YouTube-BoundingBoxes) [38] is a large data set with sufficiently long videos having an average duration of 19 seconds and 23 object categories. For DL-based trackers, the data set is split into three parts such as training, validation, and testing. All videos are manually annotated with BB from 240k unique videos.
- TrackingNet [5] is a large-scale data set having videos with an average duration of 16.6 seconds and 27 diverse object classes. Videos are taken from the YT-BB data set with the natural distribution of objects for real-time evaluation. The data set is split into 30,132 training and 511 test videos. Videos are captured at multiple resolutions and variable frame rates. Data set frames are annotated for ground truth using the VATIC tool.
- GOT-10k (Generic object tracking) [1] is a highly diversified data set with an average duration of 15 seconds and more than 560 object classes at a frame rate of 10 FPS. Out of the total 10k videos, 9.34k videos are reserved for training the model with 480 object classes and 420 are test videos with 84 object classes. The training and test video data set is exclusive without any overlapping for exhaustive testing of the trackers for practical applications. Data sets frames are manually annotated with 1.5M BB.
- RGBT234 [12] is a multi-modal large data set in which videos are recorded by both stationary and moving cameras and the longest sequence with 8k frames. RGB data is recorded by SONY EXView HAD CC camera and thermal infrared data by DLS-H37DM-A thermal camera. A two-stage annotation process is adopted to ensure the accuracy in the ground truth BB and the precise alignment between the RGB and thermal data.
- VisDrone2019 [17] consists of both diverse videos and static images. Data set is recorded by various drones such as DJI Mavic, and Phantom series (3, 3A, 3SE, 3P, 4, 4A, 4P) in diverse geographical locations from fourteen different cities of China. Videos have 10 different object categories captured at a resolution of 3840 × 2160. The data set is divided into three parts for testing, validation, and training with different videos captured in similar tracking variations. All frames are manually

annotated for ground truth. However, the ground truth for testing videos is not made public to test the real performance of the trackers.

- CDTB (Color and depth general visual object tracking benchmark) [39] is one of the largest and most diversified RGBD data sets in which indoor and outdoor sequences are recorded from multiple acquisition setups. The videos are recorded at various resolutions and multiple frame rates by the different acquisition setups. The videos from different setups are calibrated using Caltech Camera Calibration Toolbox. The sequences are manually annotated using an annotator by axis-aligned BB.
- LaSOT (Large-scale single object tracking) [16] data set has diversified videos with 70 object classes collected from YouTube. Each object class has at least 20 targets captured from natural scenes. The average duration of a video is 84 seconds which varies between 33 seconds to 6.3 minutes. All frames are manually annotated. However, the target is not annotated in case of full occlusion and out-of-view. The data set is split videos into training and testing as 1120 and 280, respectively.
- LSOTB-TIR ((Large-scale thermal infrared object tracking benchmark) [40] is a highly diversified data set containing thermal and IR sequences split into two parts, namely short-term tracking and long-term tracking. Four different scenarios containing 12 attributed challenges for the tracker's tedious performance evaluation. Manual annotation of the object in each frame within a total 770k BB.
- DTTD (Digital Twin Tracking Data set) [41] is a multi-modal data set having videos from depth sensors and vision sensors. The videos contain various occlusion scenes and varying lighting conditions. The data set is suitable to analyze the tracker's performance for long-range applications.

5.5 STRENGTHS AND LIMITATIONS OF PUBLIC TRACKING DATA SETS

With the progress in tracking algorithms, public data sets are evolved to evaluate the tracker's performance in real time. It has been well acknowledged that tracking data sets should be effective and efficient enough to analyze the tracker's accuracy and robustness before practical deployment. There are lots of data sets published by many researchers every year [19]. The developer should be capable enough in choosing the potential data set as per the tracker's appearance model design and architecture which can analyze the tracker's performance critically.

Initially, data sets have been proposed which contain a pool of videos of shorter duration [6]. Shorter videos recorded with visual tracking challenges are considered suitable enough to capture the statistics of the tracking algorithms. However, shorter videos are not able to incorporate multiple attributed challenges in a single video. After that, data sets have been developed which have a sufficient number of videos of enough duration with multiple attributed challenges to evaluate the tracker's efficiency for single object tracking [2, 9]. These data sets have gained popularity among researchers due to the wide range of videos and diverse backgrounds along with annotated ground truth for quantitative evaluations. But these data sets have a limited number of videos that can evaluate the tracker during full occlusion, out-of-view, and small-size target challenges. Also, most of the videos are recorded with a static camera only. There is no aerial view video included in these data sets that can estimate trackers for search/rescue operations, troops tracking and/or crowd monitoring. Further, aerial view videos are recorded with moving cameras mounted on UAVs captured from low altitude and presented by [4, 24, 25]. These data sets have videos captured at a wider angle with small-size objects, abrupt scale variations, ARC, and low resolution. The videos are suitable for the comprehensive analysis of the trackers with multiple pose variations along with fast and rotational challenges in realistic scenarios.

To expand the capability of trackers during day and night, multi-modal information extracted from different sensors can be integrated into the tracker's appearance model [42, 43]. To evaluate the tracker in varying lighting, extreme weather, and night conditions, multi-sensor data such as depth, thermal, and IR is integrated with RGB data. Multi-modal data sets have been proposed including the depth and thermal information with the RGB. Multi-modal data sets are available as limited size [32, 36] as well as large data sets [12, 39] to cater to the requirements of trackers. Limited-size data sets contain a handful of videos which are insufficient for the overall evaluation of the trackers. However, these videos can analyze the behavior of the multi-modal trackers as they have integrated the information from different modalities. On the other hand, large data sets contain a large number of videos with a vast variety of real-time challenges. But large numbers of videos require intensive computations for testing tracker performance. Also, these data sets contain video taken from the stationary camera as well as moving the camera to deeply analyze the tracking failures. The alignment between the vision data and specialized sensor data is highly important to obtain accurate and unbiased tracking results.

With significant progress in the tracking domain, DL-based trackers have been proposed due to their accuracy and potential to handle tedious environmental conditions. DL-based trackers need a lot of data for training [16]. Also, it has been analyzed for real-time applications of trackers, performance needs to be evaluated on long-term data sets. Long-term data sets not only include a large set of videos but also the average duration of the video is quite long. Long data sets focus to test the tracker's efficiency in real-time scenarios due to diversified videos with various object classes and variable background situations. Most videos in these data sets include wildlife scenarios that have multiple and complex attributed challenges. These data sets specify the data set split into training, testing, and validation to ensure the non-overlapping among these sets and fair evaluation of the tracker. Long-term data sets gauge the gap between the increasing numbers of deep learning trackers and the deficiency of data sets with a large number of videos with lots of object classes along with diverse background scenarios.

5.6 SUMMARY

Tracking benchmark plays a significant role in the fair evaluation of the tracker's performance. Benchmarks can be categorized into self-generated data sets and public data sets. Self-generated data sets are generated by the researchers for the evaluation of their algorithms only. The details of the videos under these data sets are not publicly available and hence cannot be assessed for quality analysis. On the other hand, public data sets provide a common and unified base for tracker evaluation and comparison. Videos in these data sets provide diversified and complex real-time situations for the exhaustive tracker's evaluation.

To satisfy the need and requirements of the trackers, a wide range of public benchmarks is available. For short-term tracking, data sets with smaller duration are available. Shorter videos with multiple attributed challenges have proved efficient and effective to analyze the true performance of traditional tracking solutions. For trackers exploiting multiple modalities, various RGBT, RGBD, and RGB-D-T public data sets are developed for comprehensive performance comparison. The main focus of these data sets is to record videos in complex weather conditions such as hot summer, cold winter, strong lighting, and heavy rain. The main concern with these data sets is to ensure the proper alignment of the vision sensor data with the other sensor's data for unbiased and accurate evaluation. Another popular category of public tracking data sets is large data sets with longer-duration videos. These data sets are suitable to test the trackers for long-term tracking challenges. Also, the size of the training set videos is sufficient for the data requirements of DL-based trackers. These data sets can test the real-time performance of the trackers before deployment. However, the execution of trackers on a large data set is quite costly in terms of the computational time needed to generate results.

REFERENCES

1. Huang, L., X. Zhao, and K. Huang, Got-10k: A large high-diversity benchmark for generic object tracking in the wild. *IEEE Transactions on Pattern Analysis and Machine Intelligence*, 2019. **43**(5): pp. 1562–1577.
2. Wu, Y., J. Lim, and M. Yang, Visual tracking benchmark. *IEEE Transactions on Pattern Analysis and Machine Intelligence*, 2015. **37**(9): pp. 1834–1848.

3. Kumar, A., G.S. Walia, and K. Sharma. Real-time multi-cue object tracking: Benchmark. in *Proceedings of International Conference on IoT Inclusive Life (ICIIL 2019), NITTTR Chandigarh, India*. 2020. Springer.
4. Mueller, M., N. Smith, and B. Ghanem. A benchmark and simulator for uav tracking. in *European Conference on Computer Vision*. 2016. Springer.
5. Muller, M., A. Bibi, S. Giancola, S. Alsubaihi, and B. Ghanem. Trackingnet: A large-scale data set and benchmark for object tracking in the wild. in *Proceedings of the European Conference on Computer Vision (ECCV)*. 2018.
6. Smeulders, A.W., D.M. Chu, R. Cucchiara, S. Calderara, A. Dehghan, and M. Shah, Visual tracking: An experimental survey. *IEEE Transactions on Pattern Analysis and Machine Intelligence*, 2013. **36**(7): pp. 1442–1468.
7. Walia, G.S. and R. Kapoor, Intelligent video target tracking using an evolutionary particle filter based upon improved cuckoo search. *Expert Systems with Applications*, 2014. **41**(14): pp. 6315–6326.
8. Kumar, A., G.S. Walia, and K. Sharma, Recent trends in multicue based visual tracking: A review. *Expert Systems with Applications*, 2020. **162**: p. 113711.
9. Liang, P., E. Blasch, and H. Ling, Encoding color information for visual tracking: Algorithms and benchmark. *IEEE Transactions on Image Processing*, 2015. **24**(12): pp. 5630–5644.
10. Li, C., H. Cheng, S. Hu, X. Liu, J. Tang, and L. Lin, Learning collaborative sparse representation for grayscale-thermal tracking. *IEEE Transactions on Image Processing*, 2016. **25**(12): pp. 5743–5756.
11. Patino, L., T. Cane, A. Vallee, and J. Ferryman. Pets 2016: Data set and challenge. in *Proceedings of the IEEE Conference on Computer Vision and Pattern Recognition Workshops*. 2016.
12. Li, C., X. Liang, Y. Lu, N. Zhao, and J. Tang, RGB-T object tracking: Benchmark and baseline. *Pattern Recognition*, 2019. **96**: p. 106977.
13. Li, C., N. Zhao, Y. Lu, C. Zhu, and J. Tang. Weighted sparse representation regularized graph learning for RGB-T object tracking. in *Proceedings of the 25th ACM International Conference on Multimedia*. 2017.
14. Moyà-Alcover, G., A. Elgammal, A. Jaume-i-Capó, and J. Varona, Modeling depth for nonparametric foreground segmentation using RGBD devices. *Pattern Recognition Letters*, 2017. **96**: pp. 76–85.
15. Xiao, J., R. Stolkin, Y. Gao, and A. Leonardis, Robust fusion of color and depth data for RGB-D target tracking using adaptive range-invariant depth models and spatio-temporal consistency constraints. *IEEE Transactions on Cybernetics*, 2018. **48**(8): pp. 2485–2499.
16. Fan, H., H. Bai, L. Lin, F. Yang, P. Chu, G. Deng, ... Y. Xu, Lasot: A high-quality large-scale single object tracking benchmark. *International Journal of Computer Vision*, 2021. **129**(2): pp. 439–461.
17. Zhu, P., L. Wen, X. Bian, H. Ling, and Q. Hu, Vision meets drones: A challenge. arXiv preprint arXiv:1804.07437, 2018.
18. Gao, C., Y. Du, J. Liu, J. Lv, L. Yang, D. Meng, and A.G. Hauptmann, Infar data set: Infrared action recognition at different times. *Neurocomputing*, 2016. **212**: pp. 36–47.
19. Dubuisson, S. and C. Gonzales, A survey of data sets for visual tracking. *Machine Vision and Applications*, 2016. **27**(1): pp. 23–52.

20. Oh, S., A. Hoogs, A. Perera, N. Cuntoor, C.-C. Chen, J.T. Lee, ... L. Davis. A large-scale benchmark data set for event recognition in surveillance video. in *CVPR 2011*. 2011. IEEE.
21. Kiani Galoogahi, H., A. Fagg, C. Huang, D. Ramanan, and S. Lucey. Need for speed: A benchmark for higher frame rate object tracking. in *Proceedings of the IEEE International Conference on Computer Vision*. 2017.
22. Li, S. and D.-Y. Yeung. Visual object tracking for unmanned aerial vehicles: A benchmark and new motion models. in *Thirty-First AAAI Conference on Artificial Intelligence*. 2017.
23. Kristan, M., A. Leonardis, J. Matas, M. Felsberg, R. Pflugfelder, L. Čehovin Zajc, ... A. Eldesokey. The sixth visual object tracking vot2018 challenge results. in *Proceedings of the European Conference on Computer Vision (ECCV) Workshops*. 2018.
24. Du, D., Y. Qi, H. Yu, Y. Yang, K. Duan, G. Li, ... Q. Tian. The unmanned aerial vehicle benchmark: Object detection and tracking. in *Proceedings of the European Conference on Computer Vision (ECCV)*. 2018.
25. Yu, H., G. Li, W. Zhang, Q. Huang, D. Du, Q. Tian, and N. Sebe, The unmanned aerial vehicle benchmark: Object detection, tracking and baseline. *International Journal of Computer Vision*, 2020. **128**(5): pp. 1141–1159.
26. Lukežič, A., L.Č. Zajc, T. Vojíř, J. Matas, and M. Kristan, Now you see me: Evaluating performance in long-term visual tracking. arXiv preprint arXiv:1804.07056, 2018.
27. Kalal, Z., K. Mikolajczyk, and J. Matas, Tracking-learning-detection. *IEEE Transactions on Pattern Analysis and Machine Intelligence*, 2011. **34**(7): pp. 1409–1422.
28. Čehovin Zajc, L., A. Lukežič, A. Leonardis, and M. Kristan. Beyond standard benchmarks: Parameterizing performance evaluation in visual object tracking. in *Proceedings of the IEEE International Conference on Computer Vision*. 2017.
29. Li, C., X. Sun, X. Wang, L. Zhang, and J. Tang, Grayscale-thermal object tracking via multitask Laplacian sparse representation. *IEEE Transactions on Systems, Man, and Cybernetics: Systems*, 2017. **47**(4): pp. 673–681.
30. Yan, S., J. Yang, J. Käpylä, F. Zheng, A. Leonardis, and J.-K. Kämäräinen. Depthtrack: Unveiling the power of rgbd tracking. in *Proceedings of the IEEE/CVF International Conference on Computer Vision*. 2021.
31. Lathoud, G., J.-M. Odobez, and D. Gatica-Perez. AV16.3: An audio-visual corpus for speaker localization and tracking. in *International Workshop on Machine Learning for Multimodal Interaction*. 2004. Springer.
32. Song, S. and J. Xiao. Tracking revisited using RGBD camera: Unified benchmark and baselines. in *Proceedings of the IEEE International Conference on Computer Vision*. 2013.
33. Wu, Z., N. Fuller, D. Theriault, and M. Betke. A thermal infrared video benchmark for visual analysis. in *Proceedings of the IEEE Conference on Computer Vision and Pattern Recognition Workshops*. 2014.
34. Hwang, S., J. Park, N. Kim, Y. Choi, and I. So Kweon. Multispectral pedestrian detection: Benchmark data set and baseline. in *Proceedings of the IEEE Conference on Computer Vision and Pattern Recognition*. 2015.

35. Felsberg, M., M. Kristan, J. Matas, A. Leonardis, R. Pflugfelder, G. Häger, ... Z. He. *The Thermal Infrared Visual Object Tracking VOT-TIR2016 Challenge Results*. 2016. Springer International Publishing.
36. González, A., Z. Fang, Y. Socarras, J. Serrat, D. Vázquez, J. Xu, and A.M. López, Pedestrian detection at day/night time with visible and FIR cameras: A comparison. *Sensors*, 2016. **16**(6): p. 820.
37. Palmero, C., A. Clapés, C. Bahnsen, A. Møgelmose, T.B. Moeslund, and S. Escalera, Multi-modal rgb–depth–thermal human body segmentation. *International Journal of Computer Vision*, 2016. **118**(2): pp. 217–239.
38. Real, E., J. Shlens, S. Mazzocchi, X. Pan, and V. Vanhoucke. YouTube boundingboxes: A large high-precision human-annotated data set for object detection in video. in *Proceedings of the IEEE Conference on Computer Vision and Pattern Recognition*. 2017.
39. Lukežič, A., U. Kart, J. Käpylä, A. Durmush, J.-K. Kämäräinen, J. Matas, and M. Kristan. Cdtb: A color and depth visual object tracking data set and benchmark. in *Proceedings of the IEEE/CVF International Conference on Computer Vision*. 2019.
40. Liu, Q., X. Li, D. Yuan, C. Yang, X. Chang, and Z. He, LSOTB-TIR: A large-scale high-diversity thermal infrared single object tracking benchmark. *IEEE Transactions on Neural Networks and Learning Systems*, 2023.
41. Feng, W., S.Z. Zhao, C. Pan, A. Chang, Y. Chen, Z. Wang, and A.Y. Yang, Digital Twin Tracking Data set (DTTD): A new RGB+ depth 3D data set for longer-range object tracking applications. arXiv preprint arXiv:2302.05991, 2023.
42. Walia, G.S. and R. Kapoor, Robust object tracking based upon adaptive multi-cue integration for video surveillance. *Multimedia Tools and Applications*, 2016. **75**(23): pp. 15821–15847.
43. Kumar, A., R. Jain, V. A. Devi, & A. Nayyar, (Eds.). *Object Tracking Technology: Trends, Challenges, Impact, and Applications*, 2023. Springer.

Chapter 6

Conventional framework for visual tracking

Challenges and solutions

6.1 INTRODUCTION

To improve tracking performance, various methodologies are adopted by various researchers to enhance the tracker's appearance model. Conventional tracking methods are the initial approaches utilized to localize targets in complex environmental situations. Table 6.1 shows the comparative features among the three conventional tracking approaches, namely deterministic, stochastic, and discriminative. Deterministic trackers extract the target information from the background by minimizing the cost of the objective function. Under this, MS [1, 2], and part matching-based tracking [3, 4] methodologies have been proposed to design robust tracking algorithms. Also, MS is integrated with PF, Kalman filter, and radial bias function in the tracker's appearance to address its limitation [5, 6]. Deterministic trackers have simple architecture and accurate performance during tough environmental situations. Also, multi-modal information is integrated with the appearance model to improve the tracking performance during heavy occlusion [4, 7, 8].

Mostly, generative trackers search for the most similar region to the target for tracking in the subsequent frames. The appearance model of these trackers either includes subspace learning [9–12] or sparse learning [13–16]. Subspace learning-based trackers consider targets under low-dimensional space. Target subspace is learned by principal component analysis (PCA) and localizes the target with accuracy. This is the classical and most effective method of tracking in the presence of complex tracking challenges. On the other hand, sparse learning-based trackers utilize a convex optimization approach in their appearance model for the target's global representation. Target candidates are in fewer templates by the sparse representation. These trackers are computationally efficient as the number of candidate samples is reduced in the global space. Also, the target's spatial information is exploited in the local search space to aid in the tracking results.

Primarily, discriminative trackers consider tracking as a binary classification problem and discriminate targets from the background using the foreground information. In this direction, tracking by detection (TLD) [17–20] and graph-based approaches [21–24] are proposed by various researchers.

DOI: 10.1201/9781003456322-6

Table 6.1 Comparison of deterministic, generative, and discriminative tracking approaches

Tracking features	*Deterministic approach*	*Generative approach*	*Discriminative approach*
Strategy	Cost function minimization	Search for the most similar target region	Separate the foreground from the background
Algorithms	MS and its variant, and part matching	Subspace learning and sparse representation	Tracking by detection and graph-based methods
Performance	Address partial occlusion and scale variations	Illumination variations and rotational challenges	Complex occlusion and extreme background clutter
Feature fusion	✓	✓	✓
Computational complexity	Low complexity	Moderate complexity	High complexity
Tracking type	Short-term tracking	Short-term tracking	Long-term tracking
Limitations	Not able to address illumination variation and background clutter	Failed to address full occlusion and heavy background clutter	Not able to handle out-of-view and extreme rotational variations

Trackers based on the TLD strategy consider three subtasks: tracking, learning, and detection. In the first frame, the target is detected, and based on the learning strategy the target state is estimated in the video frames. TLD-based trackers are robust to various tracking challenges such as illumination, scale, partial occlusion, and background clutters. On the other hand, graph-based trackers represent the semantic and geometric relationship between the target patches as the nodes and edges of the hypergraph. Tracking is performed by updating and eliminating the invalid nodes and impacted patches either by optimization or filtering techniques. These techniques not only enhance the efficiency of the tracker, but also reduce the computational complexity.

6.2 DETERMINISTIC TRACKING APPROACH

Deterministic tracking approaches utilize MS and its variants in the appearance model [1, 5, 25, 26]. Target part matching and segmentation are also exploited by deterministic tracking algorithms [3, 4, 27]. Inspired by the simplicity, robustness, and effectiveness to represent the target independently, MS algorithms utilize multi-sensor data such as depth and thermal for better results during total occlusion and background clutters [7, 8]. The details of various MS-based tracking algorithms are shown in Table 6.2.

Table 6.2 Description of representative work under the deterministic framework

SN	*Reference*	*Year*	*Algorithm*	*Fusion*	*Summary*
1.	Yu et al. [26]	2015	MS	Multi-scale model	3D spatial histogram along with weighted background spatial histogram to suppress background information for efficient tracking.
2.	Razavi et al. [28]	2016	Enhanced MS	–	Discriminating features for capturing the spatial relationship along with adaptive target window resizing for accurate results.
3.	Xie et al. [27]	2017	Segmentation	Region clustering	Segmentation with depth information to maintain continuity during occlusion and clustering for pixel correspondences between objects of the keyframes.
4.	Dhassi and Aarab [1]	2018	MS	Weighted sum	Multiple feature fusion along with H-∞ filter for global state estimation and MS for local state association for each feature model.
5.	Parate et al. [3]	2018	Global patch-based matching	–	Hybrid approach exploiting integral channel features for global and local template updates.
6.	Xiao et al. [4]	2018	Part matching	Weighted feature fusion	Temporal consistency for the global space candidates and the part-based matching for matching global candidates with the local candidate.
7.	Rowghanian and Ansari-Asl [6]	2018	MS and radial bias function	Confidence map linear multiplication	2D correlation coefficient to determine the similarity between the target and the reference. Optimization strategy to improve memory utilization during tracking.
8.	Li et al. [7]	2018	Patch-based tracking	Weighted score fusion	Two-stage ranking strategy to mitigate inaccurate patch weight initialization and model reliability for tracker's adaptive update.
9.	Liu et al. [8]	2018	MS	Spatial context-based fusion	3D MS tracking to boost the tracker's robustness and depth information for the tracker's fast recovery after full occlusion.
10.	Medouakh et al. [25]	2018	MS	Joint weighted fusion	Discriminant features extract structural information of the target to extract target by suppressing the background information.
11.	Kanagamalligaand Vasuki [29]	2018	MS	Adaboost classifier	Feature classification along with contour information for locating the target in the current frame based on the prior frame object model.
12.	Liu and Zhong [2]	2019	MS	MS-based feature fusion	Complementary features to address the environment variations along with four-neighborhood-search methods to prevent tracking drift during occlusion.
13.	Iswanto et al. [5]	2019	MS, PF and Kalman filter	Weighted fusion	Multiple features fuse and combine by utilizing the benefits of MS, PF, and Kalman filter.
14.	Shen [30]	2021	MS	–	Extracted non-rigid target shape information from color histogram and density gradient for faster convergence of the model.

Note: The work is arranged in the ascending order of their published year.

6.2.1 Mean shift and its variant-based trackers

Under deterministic tracking, MS-based tracking algorithms are reviewed. In this direction, the authors utilize multi-scale MS to measure the similarity between the target model and the reference model [26]. Weighted background spatial histogram is used to extract the target from the background. For adaptive scale estimation, the target position is represented in the 3D image plane. In ref. [1], authors have proposed a multi-feature interactive appearance model, using color correlogram, edge orientation histogram, and LBP histogram. MS is utilized in each appearance model to extract the local state of the target. Finally, H-∞ filtering is applied to estimate the final global state of the target. Medouakh et al. [25] have represented the texture using local phase quantization (LPQ) to improve its performance during the blur challenge. To extract the target from the background effectively, the color histogram is integrated with the LPQ texture histogram using the MS algorithm. HSV color channels and LPQ texture have represented the target jointly to localize it in the video frame. Similarly, the authors have integrated HSV color space with texture features using the MS algorithm [2]. The four-neighborhood-search method is adopted to address the partial occlusion problem and improve tracking results. However, the authors have tested the performance on limited videos, and the processing involved in calculating four-neighborhood searches is also not highlighted. Shen [30] has integrated a color histogram with the MS algorithm.

To address the shape-changing issues of the non-rigid target during tracking. The density gradient is used for the fast convergence of the model for target state estimation.

Conventional MS framework-based trackers show drift during similar backgrounds and are not efficient enough to capture the spatial relationship between the target's pixels. To address these defects, PF, Kalman filter, and radial bias function are integrated with the MS algorithm for more powerful and rich target representation along with fast convergence of the tracker. For this, authors have proposed an enhance and improved MS-based tracker [28]. The interlaced derivative pattern is used to modify the texture feature to capture the spatial relationship between the target's pixels for extracting the more discriminating target's features. In order to prevent tracking failures, adaptive target window resizing is used. In ref. [29], the authors have integrated the results of two methods to obtain the target's motion information. Along with motion, shape, and Gabor features are also extracted in each frame to localize the target. Adaboost classifier is used to classify the extracted features for obtaining efficient tracking results. But the authors have utilized the radial bias function in the MS framework [6]. The 2-D correlation coefficient between the target and the reference is determined for computing the target's scale and orientation. Optimization strategy is used to minimize the tracker's computation for target localization. Fuzzy C-means clustering is used for clustering the multi-features, namely color, texture, gradient, intensity, contrast, and spatial frequency, extracted in any frame to address tracking challenges. However, to overcome the limitations of MS, PF, and Kalman filter, the authors have integrated them to propose an improved tracking framework [5]. Also, color and texture are utilized in the appearance model for improving tracking accuracy by catering to numerous tracking issues. Experimental results are not sufficient enough to prove the potential of the proposed algorithm. Integrating three frameworks also requires a lot of processing power in terms of the target's state estimation.

6.2.2 Multi-modal deterministic approach

Complementary features from multi-modal sensors are concatenated to deal with the more complex tracking challenges. In this direction, Liu et al. [8] have proposed a 3-D MS-based tracker with color and depth data for handling the severe occlusion challenge. The adaptation strategy is employed by spatial context information to choose the appropriate sample for robust tracking. But the tracker has shown drift during long-term tracking. To address the total occlusion, spatiotemporal segmentation was used to segment the target under occlusion to maintain continuity [27]. The target inconsistency after occlusion is handled by levering scale-invariant feature transform and bilateral representation. Consistent segments are generated in a few keyframes and propagated in the complete video via mask propagation scheme. Also, spatiotemporal consistency was used to achieve stability during occlusion [4]. Contextual information is integrated with depth data to prevent the false update of the tracker during occlusion. Multiple features are adaptively fused with robust re-learning of the model in the global layer. This layer is decomposed to track small target parts in the local candidate region. Li et al. [7] have represented the target as a set of patches using color and thermal features adaptively to eliminate the background noise. Each patch is provided with a weight and optimized using manifold ranking. The structured SVM is adopted to predict the target location. To summarize, multi-modal information in the MS framework is capable to address the severe tracking challenges. These trackers have demonstrated efficient and superior results in comparison to other modern equivalents. But fast processing and computation of multi-modal information demand further investigation for a strong and stable tracker.

6.3 GENERATIVE TRACKING APPROACH

Generally, generative tracking approaches are broadly categorized into subspace learning-based trackers and sparse learning-based trackers. Also, multi-modal features are extracted and integrated for a more robust tracking solution. Details about the various tracking algorithms under the generative approach are shown in Table 6.3.

Table 6.3 Description of representative work under the generative framework

SN	*Reference*	*Year*	*Algorithm*	*Update scheme*	*Summary*
1.	Zhang et al. [31]	2015	Sparse hashing	Hamming distances	Dynamic feature selection based on the sparsity of the hash coefficient vectors for the stable tracker.
2.	Sui et al. [11]	2015	Subspace learning	Mutual correlation	Apply joint row-wise sparsity to obtain the mutual relationship between the observed features.
3.	Li et al. [32]	2017	Template matching	Online template update	Perform both online and offline template matching to localize the target in the current frame.
4.	Feng et al. [13]	2017	Sparse representation	Similarity calculation	Compute similarity based on the temporal context information from the group of prior target templates.
5.	Hu et al. [33]	2017	Template matching	Confidence score	Fuse information from two channels of the binocular camera using epipolar geometry.
6.	Jansi et al. [14]	2017	Sparse learning	Structural similarity	Calculate Pearson's correlation to generate representative dictionary to update the appearance model adaptively.
7.	Ma and Xiang [34]	2017	Sparse learning	Depth-based occlusion detection	Occlusion detection along with depth-based occlusion detection to address occlusion issues efficiently.
8.	Li et al. [35]	2017	Sparse representation	Weighted update	Adaptively fuse multi-modal data in a Laplacian sparse representation with model reliability.
9.	Wang et al. [36]	2018	Joint sparse representation	ADMM	Exploit the underlying relationship between features to select the robust features for target representation.
10.	Sui et al. [10]	2018	Subspace learning	Low-rank approximation	Represent the target and the neighboring background using subspace learning.
11.	Lan et al. [37]	2018	Sparse learning	Online multiple metric learning	Multiple sparse learning with feature-specific properties for the informative tracker's appearance model.
12.	Kang et al. [38]	2019	Sparse representation	Discriminant matrix	To enhance robustness, multi-view sparse along with non-local regularizer with spatial smoothness.
13.	Li et al. [9]	2019	Incremental subspace learning	Selection based	Select the most reliable tracker results for the target's final state estimation.
14.	Javanmardi and Qi [15]	2019	Sparse learning	ADMM	Spatial structure of the candidate samples by employing group sparsity regularization.
15.	Tian et al. [16]	2020	Sparse learning	Alternating iteration	Low-rank constraint in order to eliminate similar particles to reduce the computational complexity.
16.	Tai et al. [12]	2021	Subspace reconstruction	Dynamic L1-PCA	Image patch reconstruction during abrupt environmental variation to prevent drift.

Note: The work is arranged in the ascending order of their published year.

6.3.1 Subspace learning-based trackers

Subspace learning-based trackers are recently explored under the generative tracking framework [9–12]. In ref. [9], the authors propose a tracker consisting of two appearance models using two algorithms, namely Bayesian and incremental subspace learning. Incremental sub-learning computes the target's subspace coordinates based on the maximum posterior probability and a Bayesian classifier is used to select reliable candidates for target state estimation with online learning. The target and the surrounding background was constructed using subspace learning [10]. Discriminative low ranking classifies unreliable samples to predict the correct target labels. Subspace separates the target from the background linearly. However, sparsity was induced in the subspace learning to reduce the accumulated errors due to distractive target information in the appearance model [11]. The method also exploits the relationship between the features during learning, hence increasing the computations for real-time tracking. Experimental results demonstrate the robustness of the method but on the limited number of video sequences. Tai et al. [12] replace the original image patch with the reconstructed subspace during filter learning to handle the abrupt target variations. L1-PCA determines the outlier and updates the subspace with effective information. In summary, subspace methods are efficient in tracking the target in complex environmental variations. These methods employ a robust update strategy to avoid the erroneous update of the tracker with contaminated samples.

6.3.2 Sparse representation-based trackers

Sparse learning-based trackers generate strong discriminative features capable to prevent the tracker's drift during abrupt deformation in the target's appearance. In this direction, the authors have combined the target's spatial context and temporal context information using sparse representation [13]. Target similarity between the consecutive frames is captured by using an image quality assessment method based on the spectral residual. Multiple historical templates are generated from prior tracking results to calculate the similarity with the target in the current frame. But Jansi et al. [14] propose a representative dictionary learning using the sparse method to determine the similarity between the target in the current frame and the dictionary. An appearance model structural similarity test based on the simple binary mask is also performed to identify the target among the various other candidates in the scene. In ref. [15], the authors have exploited local spatial information along with group sparsity regularization terms to represent the target candidate. Target spatial information is preserved by the convex optimization model and the model is converged to a solution iteratively by the alternating direction method of multiplier (ADMM). Similarly, the authors have utilized ADMM to update the model with a small number of candidate samples using group sparsity [36]. Multiple features, namely color, texture, and edge are sparsely represented and linearly fused in the adaptive appearance model. Online feature selection method is employed to obtain the robust features from the samples of the target and the background. Similarly, authors have used multi-feature sparse representation along with an online regularizer with multiple metric learning [37].

The target appearance model is built from the feature-specific learning from the common and complementary features. Kang et al. [38] have proposed a multi-view sparse representation method to extract useful information from unreliable features. Group projection from both observation groups, i.e. reliable and unreliable features, is exploited to obtain the group similarity using a discriminant matrix. Spatial smoothness among the different groups, i.e. local as well as global is enforced using a non-local regularizer. However, the authors have employed a fraction order regularizer between the consecutive frames to address the environmental variations during model updates [16]. Convex low-rank constraint along with inverse sparse representation is used to model the similarity relationship between the target and the candidate samples. Zhang et al. [31] have used hash function to model the target and candidate templates in a sparse framework. Both inter-class and intra-class relationships of the target template are exploited to improve the classification. Hash coefficient vectors are used for dynamic and discriminative feature selection during tracking variations. On the other hand, the authors have performed adaptive multi-feature template matching by adaptively adjusting the feature weight for the target template during variations [32]. The target is located using both online and offline template matching. In summary, the similarity between the target and the candidate samples is determined using various sparse representation-based generative tracking to track the target efficiently during complex occlusion and severe deformations.

6.3.3 Multi-modal generative approach for visual tracking

Sparse trackers based on multi-modal data have been widely explored to address varying illumination, extreme occlusion, and similar background [33, 34]. These trackers provide more discriminative feature information suitable to prevent the tracker's drift. In this direction, authors have integrated depth data with the color information in a sparse framework [34]. An occlusion template set is generated from the segmented occlusion region from the target area to handle complex occlusion. Depth-based histogram occlusion detection strategy is used to determine the correct time of template update for effective tracking. In ref. [33], the authors have utilized binocular camera information for long-term tracking. The complementary information from two channels using epipolar geometry is exploited to estimate the target's location. Binocular information is robust enough to address the illumination change, viewpoint change, and scale variations effectively. Li et al.

[35] fuse the multi-modal data, viz. grayscale and thermal, in a Laplacian sparse framework adaptively. The similarity between the patch pairs and the reliability of the patch weights is computed to measure the confidence score. This score is extracted to build the relationship for the target patch for efficient tracking. In sum, multi-modal information in the generative model is not only efficient against extreme environmental conditions but also reduces the computational processing for target tracking.

6.4 DISCRIMINATIVE TRACKING APPROACH

Discriminative trackers have shown superior performance during severe occlusion and background clutter as they have exploited algorithms to separate the target from the background. These trackers have robust performance in comparison to other tracking algorithms and are categorized either as tracking by the detection or graph-based trackers. The salient features of the discriminative tracker under these categories are shown in Table 6.4.

6.4.1 Tracking by detection

In this category of discriminative trackers, tracking is performed in three steps: tracking, learning, and detection (TLD). Primarily, algorithms based on TLD methodology have shown efficient performance during long-term tracking [17, 19]. In this direction, the target's texture information was extracted by exploiting LBP and BB near to tracking a target using the nearest neighbor classifier [19]. Target detection is performed using three classifiers: filter, ensemble, and variance. The learning task is to detect the target in the first frame and update the detector as per the tracking in video frames. Tracking, learning, and detection are accomplished simultaneously to obtain the target state. Authors have proposed a robust tracking method based on the TLD framework using square root cubature Kalman filter to address occlusion and illumination [17]. Fast retina key points are exploited to handle rotational and scale variations. Target's BB detection is executed using a normalized cross-correlation coefficient. In ref. [20], the authors have modified the residual network to capture the target's appearance variations during tracking. Semantic features and discriminant features are captured from different layers of ResNet for accurate localization of the target. Dynamic update and threshold-based detection is employed when tracking failure is detected. Hu et al. [23] have addressed tracking failure and model degradation by integrating 1-D filter with 2-D filter. 1D filter estimates state while the 2D filter determines location. Also, three complementary features along with penalty-based model adaptive updates are employed to ensure the tracker's stability during tedious tracking variations. Online multiple-instance learning compares both templates, background and target, while tracking [43].

Table 6.4 Description of representative work under the discriminative framework

SN	*Reference*	*Year*	*Algorithm*	*Update strategy*	*Summary*
1.	Lu et al. [39]	2013	Locally connected graph	Bayesian method	Static learning process and Bayes-based tracking to prevent tracker drift.
2.	Jia et al. [19]	2015	Tracking by detection	Learning component-based update	LBP and nearest neighbor classifier mechanism to localize the target efficiently.
3.	Yang et al. [40]	2015	Multi-graph ranking	Incremental update	Integrate multiple graphs in regularization framework and an iterative model optimization for faster and more accurate estimation.
4.	Du et al. [41]	2016	Structural aware hypergraph	Incremental update	Structural pairwise dependencies between the target's part in multiple successive frames to discriminate it from the background.
5.	Du et al. [21]	2017	Geometric hypergraph learning	Pairwise coordinate update	Pairwise geometric relationship between local parts and high-order relationship among the correspondence hypothesis in different video frames.
6.	Wang and Ling [42]	2017	Sequential graph matching	Update based on probability distribution	Target's structural information and key point correspondence in a geometric graph to localize the target.
7.	Hu et al. [23]	2017	Tracking by detection	Adaptive update	Multi-feature fusion along with kernel correlation and interpolation for faster tracking and scale filter to address scale variations.
8.	Lu et al. [24]	2017	Probabilistic hypergraph ranking	Template update	Integrate three distinct hypergraphs linearly and determine the discriminative information using adaptive template matching.
9.	Wu et al. [43]	2018	Multiple-instance learning	Template update	Multi-feature with online weak classifier and background template update to discriminate the target from the background in a video frame.
10.	Du et al. [22]	2018	Iterative graph seeking	Incremental update	Integrate local part's structural information and global information for target state estimation.
11.	Liu et al. [20]	2019	Tracking by detection	Threshold-based dynamic update	Utilize spatial and temporal target information considering regression and classification loss to localize target accurately.
12.	Dong et al. [17]	2019	Tracking by detection	Online update	Square root cubature Kalman filter and fast retina keypoint to address occlusion, illumination, and scale variations.
13.	Zheng et al. [44]	2021	Relational graph	Online update	Local multi-patch and area spatial distance to discriminate the target from the background for extracting local features and accurate tracking.

Note: The work is arranged in the ascending order of their published year.

Reliable pixels which belong to the target are used for the adaptive update of the tracker using a predefined threshold. Handcrafted features such as HOG along with weak classifiers are exploited for the online learning of the tracker for the robust appearance model. In summary, TLD-based trackers detect the target in the first frame and keep learning the appearance model for target tracking and adaptive update during tracking variations.

6.4.2 Graph-based trackers

Tracking algorithms in this category focus on capturing the inter-relationship between the foreground and background to efficiently cope with tracking challenges. For this, the authors have integrated multiple graphs for various features in an l_2– norm regularization framework and ranked them iteratively [40]. Temporal consistency between the adjacent frames in multigraph learning to capture the target similarity in consecutive video frames. Du et al. [22] have integrated part selection, matching, and state estimation using iterative graph-seeking methodology. The energy minimization method is used to combine complementary feature information for stable tracking results. To cater to the target appearance variations, the authors have presented tracking by constructing supervised locally connected graphs [39]. Constraint connection among sub-graphs is combined using semantic subspace and Bayesian tracking in the tracker's update model. In ref. [41], the authors have captured the higher-order relationship of the target in the consecutive frames using the structure-aware hypergraph. Motion consistent target-dense graphs are extracted to determine the optimal state of the target. However, they have considered the geometric relations between the local parts of different video frames to extract the high-order relationship of the target [21]. Confidence-aware sampling is performed to scale the large network as well as to reduce the impact of noise in state estimation. The sampling strategy reduces the tracker's computation by scaling the hypergraph. Wang and Ling [42] have proposed to capture complete structural information for efficient tracking results. Planar objects are represented using key points for predicting object pose and point matching. Candidate graph construction and filter matching along with optimization are used for efficient state estimation of the target. In ref. [44], the authors have utilized patch representation and patch matching in a relational graph framework. Multi-patch representation is used to separate the target from the background and discriminative patches are constructed with a geometric layout to address tracking disturbances. Spatiotemporal-based filtering is adopted to remove the invalid patches to prevent tracking drift as well as to reduce computational complexity. To summarize, various graph-based strategies are adopted to propose robust tracking solutions. Filtering and optimization techniques are used to improve the tracker's processing capabilities [45].

6.5 SUMMARY

In this chapter, we have highlighted the salient features and limitations of three conventional tracking approaches: deterministic, generative, and discriminative. Tracking algorithms based on a deterministic approach focus on minimizing object tracking cost function to obtain efficient results. These trackers utilize MS and its variants in appearance models. However, due to the limitation of MS in handling abrupt color variations, these trackers are not able to address illumination variations, background clutter, and rotational variations efficiently. To cater to these challenges, trackers based on a generative approach are proposed. These trackers search for the most similar region to the target in the search space and perform tracking. These trackers exploit either subspace learning or sparse representation in the appearance model. Due to this, generative trackers report tracking failures during extreme occlusion, and complex background clutter in long-term videos. Discriminative trackers address these challenges efficiently in long-term videos. For this, discriminative trackers exploit tracking by detection and graph-based approach in the appearance model. These approaches aim to classify the target from the background during complex environmental variations. But the choice of classifier, tracking sample, and heavy computation model restricts the effective tracking results in real-time.

The choice of tracking model is specific to the application and the model cost. Each model has its benefits and limitations in the tracking domain. In another direction, to have accurate and efficient tracking performance, the drawbacks of generative and discriminative approaches are addressed in a collaborative model. Also, to address the high computational limitations of discriminative models, coarse-to-fine tracking methods are explored by the researchers.

REFERENCES

1. Dhassi, Y. and A. Aarab, Visual tracking based on adaptive mean shift multiple appearance models. *Pattern Recognition and Image Analysis*, 2018. **28**(3): pp. 439–449.
2. Liu, J. and X. Zhong, An object tracking method based on Mean Shift algorithm with HSV color space and texture features. *Cluster Computing*, 2019. **22**(3): pp. 6079–6090.
3. Parate, M.R., V.R. Satpute, and K.M. Bhurchandi, Global-patch-hybrid template-based arbitrary object tracking with integral channel features. *Applied Intelligence*, 2018. **48**(2): pp. 300–314.
4. Xiao, J., R. Stolkin, Y. Gao, and A. Leonardis, Robust fusion of color and depth data for RGB-D target tracking using adaptive range-invariant depth models and spatio-temporal consistency constraints. *IEEE Transactions on Cybernetics*, 2018. **48**(8): pp. 2485–2499.

5. Iswanto, I.A., T.W. Choa, and B. Li, Object tracking based on mean shift and particle-Kalman filter algorithm with multi features. *Procedia Computer Science*, 2019. **157**: pp. 521–529.
6. Rowghanian, V. and K. Ansari-Asl, Object tracking by mean shift and radial basis function neural networks. *Journal of Real-Time Image Processing*, 2018. **15**(4): pp. 799–816.
7. Li, C., C. Zhu, S. Zheng, B. Luo, and J. Tang, Two-stage modality-graphs regularized manifold ranking for RGB-T tracking. *Signal Processing: Image Communication*, 2018. **68**: pp. 207–217.
8. Liu, Y., X.-Y. Jing, J. Nie, H. Gao, J. Liu, and G.-P. Jiang, Context-aware three-dimensional mean shift with occlusion handling for robust object tracking in RGB-D videos. *IEEE Transactions on Multimedia*, 2018. **21**(3): pp. 664–677.
9. Li, K., F. He, H. Yu, and X. Chen, A parallel and robust object tracking approach synthesizing adaptive Bayesian learning and improved incremental subspace learning. *Frontiers of Computer Science*, 2019. **13**(5): pp. 1116–1135.
10. Sui, Y., Y. Tang, L. Zhang, and G. Wang, Visual tracking via subspace learning: A discriminative approach. *International Journal of Computer Vision*, 2018. **126**(5): pp. 515–536.
11. Sui, Y., S. Zhang, and L. Zhang, Robust visual tracking via sparsity-induced subspace learning. *IEEE Transactions on Image Processing*, 2015. **24**(12): pp. 4686–4700.
12. Tai, Y., Y. Tan, S. Xiong, and J. Tian, Subspace reconstruction-based correlation filter for object tracking. *Computer Vision and Image Understanding*, 2021. **212**: p. 103272.
13. Feng, P., C. Xu, Z. Zhao, F. Liu, C. Yuan, T. Wang, and K. Duan, Sparse representation combined with context information for visual tracking. *Neurocomputing*, 2017. **225**: pp. 92–102.
14. Jansi, R., R. Amutha, R. Alice, E.M. Chitra, and G.S. Ros. Robust object tracking using sparse-based representative dictionary learning. in *2017 International Conference on Computation of Power, Energy Information and Communication (ICCPEIC)*. 2017. IEEE.
15. Javanmardi, M. and X. Qi, Structured group local sparse tracker. *IET Image Processing*, 2019. **13**(8): pp. 1391–1399.
16. Tian, D., G. Zhang, and S. Zang, Robust object tracking via reverse low-rank sparse learning and fractional-order variation regularization. *Mathematical Problems in Engineering*, 2020. pp. 1–10.
17. Dong, E., M. Deng, J. Tong, C. Jia, and S. Du, Moving vehicle tracking based on improved tracking–learning–detection algorithm. *IET Computer Vision*, 2019. **13**(8): pp. 730–741.
18. Hare, S., S. Golodetz, A. Saffari, V. Vineet, M.-M. Cheng, S.L. Hicks, and P.H. Torr, Struck: Structured output tracking with kernels. *IEEE Transactions on Pattern Analysis and Machine Intelligence*, 2015. **38**(10): pp. 2096–2109.
19. Jia, C., Z. Wang, X. Wu, B. Cai, Z. Huang, G. Wang, … D. Tong. A Tracking–Learning–Detection (TLD) method with local binary pattern improved. in *2015 IEEE International Conference on Robotics and Biomimetics (ROBIO)*. 2015. IEEE.
20. Liu, B., Q. Liu, T. Zhang, and Y. Yang, MSSTResNet-TLD: A robust tracking method based on tracking–learning–detection framework by using multi-scale

spatio-temporal residual network feature model. *Neurocomputing*, 2019. **362**: pp. 175–194.
21. Du, D., H. Qi, L. Wen, Q. Tian, Q. Huang, and S. Lyu, Geometric hypergraph learning for visual tracking. *IEEE Transactions on Cybernetics*, 2017. **47**(12): pp. 4182–4195.
22. Du, D., L. Wen, H. Qi, Q. Huang, Q. Tian, and S. Lyu, Iterative graph seeking for object tracking. *IEEE Transactions on Image Processing*, 2018. **27**(4): pp. 1809–1821.
23. Hu, Q., Y. Guo, Z. Lin, W. An, and H. Cheng, Object tracking using multiple features and adaptive model updating. *IEEE Transactions on Instrumentation and Measurement*, 2017. **66**(11): pp. 2882–2897.
24. Lu, R., W. Xu, Y. Zheng, and X. Huang, Visual tracking via probabilistic hypergraph ranking. *IEEE Transactions on Circuits and Systems for Video Technology*, 2017. **27**(4): pp. 866–879.
25. Medouakh, S., M. Boumehraz, and N. Terki, Improved object tracking via joint color-LPQ texture histogram based mean shift algorithm. *Signal, Image and Video Processing*, 2018. **12**(3): pp. 583–590.
26. Yu, W., X. Tian, Z. Hou, Y. Zha, and Y. Yang, Multi-scale mean shift tracking. *IET Computer Vision*, 2015. **9**(1): pp. 110–123.
27. Xie, Q., O. Remil, Y. Guo, M. Wang, M. Wei, and J. Wang, Object detection and tracking under occlusion for object-level RGB-D video segmentation. *IEEE Transactions on Multimedia*, 2017. **20**(3): pp. 580–592.
28. Fatemeh Razavi, S., H. Sajedi, and M. Ebrahim Shiri, Integration of colour and uniform interlaced derivative patterns for object tracking. *IET Image Processing*, 2016. **10**(5): pp. 381–390.
29. Kanagamalliga, S. and S. Vasuki, Contour-based object tracking in video scenes through optical flow and gabor features. *Optik*, 2018. **157**: pp. 787–797.
30. Shen, Z. Sports video tracking technology based on mean shift and color histogram algorithm. in *Journal of Physics: Conference Series*. 2021. IOP Publishing.
31. Zhang, L., H. Lu, D. Du, and L. Liu, Sparse hashing tracking. *IEEE Transactions on Image Processing*, 2015. **25**(2): pp. 840–849.
32. Li, Z., S. Gao, and K. Nai, Robust object tracking based on adaptive templates matching via the fusion of multiple features. *Journal of Visual Communication and Image Representation*, 2017. **44**: pp. 1–20.
33. Hu, M., Z. Liu, J. Zhang, and G. Zhang, Robust object tracking via multi-cue fusion. *Signal Processing*, 2017. **139**: pp. 86–95.
34. Ma, Z.-A. and Z.-Y. Xiang, Robust object tracking with RGBD-based sparse learning. *Frontiers of Information Technology & Electronic Engineering*, 2017. **18**(7): pp. 989–1001.
35. Li, C., X. Sun, X. Wang, L. Zhang, and J. Tang, Grayscale-thermal object tracking via multitask Laplacian sparse representation. *IEEE Transactions on Systems, Man, and Cybernetics: Systems*, 2017. **47**(4): pp. 673–681.
36. Wang, Y., X. Luo, L. Ding, and S. Hu, Visual tracking via robust multi-task multi-feature joint sparse representation. *Multimedia Tools and Applications*, 2018. **77**(23): pp. 31447–31467.
37. Lan, X., S. Zhang, P.C. Yuen, and R. Chellappa, Learning common and feature-specific patterns: A novel multiple-sparse-representation-based tracker. *IEEE Transactions on Image Processing*, 2018. **27**(4): pp. 2022–2037.

38. Kang, B., W.-P. Zhu, D. Liang, and M. Chen, Robust visual tracking via non-local regularized multi-view sparse representation. *Pattern Recognition*, 2019. **88**: pp. 75–89.
39. Lu, K., Z. Ding, and S. Ge, Locally connected graph for visual tracking. *Neurocomputing*, 2013. **120**: pp. 45–53.
40. Yang, X., M. Wang, and D. Tao, Robust visual tracking via multi-graph ranking. *Neurocomputing*, 2015. **159**: pp. 35–43.
41. Du, D., H. Qi, W. Li, L. Wen, Q. Huang, and S. Lyu, Online deformable object tracking based on structure-aware hyper-graph. *IEEE Transactions on Image Processing*, 2016. **25**(8): pp. 3572–3584.
42. Wang, T. and H. Ling, Gracker: A graph-based planar object tracker. *IEEE Transactions on Pattern Analysis and Machine Intelligence*, 2017. **40**(6): pp. 1494–1501.
43. Wu, F., S. Peng, J. Zhou, Q. Liu, and X. Xie, Object tracking via online multiple-instance learning with reliable components. *Computer Vision and Image Understanding*, 2018. **172**: pp. 25–36.
44. Zheng, J., Y. Xu, and M. Xin, Structured object tracking with discriminative patch attributed relational graph. *Knowledge-Based Systems*, 2021. **225**: p. 107097.
45. Kumar, A., G.S. Walia, and K. Sharma (2020). Recent trends in multicue based visual tracking: A review. *Expert Systems with Applications*, **162**: p. 113711.

Chapter 7

Stochastic framework for visual tracking

Challenges and solutions

7.1 INTRODUCTION

PF is used extensively to provide efficient tracking solutions due to ease of implementation and accuracy in results [1, 2]. It utilizes sequential Monte Carlo simulations in the Bayesian framework for the target's state estimation. PF consists of two steps namely, the prediction step and the likelihood calculation step for state estimation [3]. During the prediction step, particles are evolved using the state model and the likelihood calculation step determines the particle's weight during the tracker's updated model. Table 7.1 shows the potential work under the stochastic framework along with their salient features.

PF offers advantages in comparison to its other variants to handle tough environmental variations effectively. In this direction, authors have proposed spider monkey optimized PF to improve the tracking efficiency in the PF framework [4]. Optimization not only enhances the quality of the particles but also improves the diversity in the search space. Color is extracted in the first frame of the video as a feature for each particle on the target. The particle weight is updated using the global leader phase of the meta-heuristics optimization. Similarly, authors have utilized a color histogram for each particle on the target [5]. New particle weights are initialized using levy flight and distributed in the search space using improved cuckoo search algorithms. The particles are resampled only when their weights fall below a predefined threshold. However, the authors have proposed to optimize the particles using the firefly algorithm to move them to the high-likelihood region before resampling them [6].

Truong et al. [9] have proposed to calculate CH using a saliency-based weighting scheme to address its limitations during background clutter. Particles are initialized on the target and are resampled at the end of the iteration. The newly generated particles are redistributed in the search space to localize the target effectively. On the other hand, Zhang et al. [12] have extracted the gradient information from the particles initialized on the target. Particles are resampled using hummingbird optimization. The optimal value of the optimization's fitness function represents the optimal solution

DOI: 10.1201/9781003456322-7

Table 7.1 Description of representative work under the stochastic framework

SN	*References*	*Year*	*Feature extracted*	*Resampling methods*	*Number of particles*	*Summary*
1.	Walia et al. [5]	2014	Color	Levy flight-based Improved cuckoo search optimization	100	Handled target rotational and scale variations using optimization in the PF framework.
2.	Gao et al. [6]	2015	Color	Firefly optimization	10 and 20	Optimized particles before the resampling techniques to ensure particles in high-likelihood regions.
3.	Xiao et al. [7]	2016	Color and HOG	Conventional PF resampling	—	Weighted adaptive fusion of features to address the tracking challenges efficiently.
4.	Jiang et al. [8]	2016	Intensity, HOG, and CN	—	300	Online adaptive fusion of confidence maps of each feature to obtain the final map for each particle.
5.	Rohilla et al. [4]	2017	Color	Spider monkey optimization	150	Meta-heuristic optimization-based resampling ensured diversity in the search space.
6.	Truong et al. [9]	2018	Weighted CH	—	300	Preserved spatial information by the high-weighted particles in the pixel.
7.	Walia et al. [10]	2018	Color, edge, and LBP	Cuckoo search optimization	40	Determined the final particle's weight using the DSmt theory-based PCR-6 fusion method.
8.	Cai-xia and Xin-yan [11]	2019	CH and HOG	Sequential importance resampling	—	Used fuzzy C-means clustering to reduce the number of particles for tracking.
9.	Zhang et al. [12]	2019	Gradient	Hummingbirds optimization	100, 500 and 1000	High number of particles is used for state estimation exploiting a single feature in the appearance model.
10.	Narayana et al. [1]	2019	Color	Modified grey wolf optimization	100	Generated new particles and resampled all using the resampling technique.
11.	Xue and Li [2]	2019	Local sparse	Spatio-temporal based confidence map	200	Spatio-temporal context information for resampling of the target's confidence maps.

(*Continued*)

Table 7.1 (Continued)

SN	*References*	*Year*	*Feature extracted*	*Resampling methods*	*Number of particles*	*Summary*
12.	Moghaddasi and Faraji [13]	2020	HSV Color and Texture	Genetic algorithm	20–60	Particles evolved by the resampling technique to address PF limitations.
13.	Dash and Patra [14]	2020	Ohta color model and texture	MS optimization	10	Utilized quantum-based PF to address abrupt motion variations.
14.	Wang et al. [15]	2020	Locality sensitive histogram	Ant colony optimization	200	Performed and compared experimental results on limited videos.
15.	Kumar et al. [16]	2020	Color, Texture, and PHOG	Butterfly search optimization	50	Utilized resampling technique after the fusion of the particles to ensure diversity.
16.	Walia et al. [17]	2020	Color and LBP	Crow search optimization	50	Outlier detection mechanism reduced the number of processing particles for the resampling technique.
17.	Kiliç et al. [18]	2014	HSV and audio	Sampling importance resampling	10	Method used for speaker tracking and finding the optimal number of particles for efficient tracking.
18.	Xiao et al. [19]	2016	Color and IR	Importance sampling	—	Integrated the target tracking using color and template matching using IR data along with occlusion handling.
19.	Walia et al. [20]	2016	Color, motion, and thermal	Cuckoo search	40	Adaptive multi-feature fusion using PCR5 for enhancement and suppression of particles.
20.	Zhang et al. [21]	2017	Sparse and IR	Cumulative probability distribution sampling	50	The proposed method for tracking small objects distinguishes the target from the background.

for the tracker. However, the effective number of used particles is quite high. Color information was extracted from the particles and meta-heuristics optimization as a resampling technique to reduce the effective number of particles for target tracking [1]. The particles are optimized before the resampling technique to expand the efficient particle for precise estimation of the target's actual location. Also, the authors have proposed to extract the target's local and structural information by using a sparse-based appearance model [2]. Spatio-temporal context information-based resampling is used to diversify the particles in the search space. The multi-task learning strategy for particle learning is adopted to improve tracking results. Wang et al. [15] have utilized ant colony optimization to move particles in high-likelihood regions. An unscented Kalman filter is used to predict the better particle state for the target's precise localization. However, the method uses a high number of particles, and processing such a number of particles is computationally complex. In summary, the number of particles should be reduced for the tracker's efficient performance in terms of computation. However, accurate state prediction with fewer particles is challenging. In addition, sufficient features should be incorporated to address the real-time tracking challenges.

Single feature-based trackers are not able to address the dynamic real-time tracking challenges efficiently. Hence, multi-features-based trackers have gained popularity due to efficient tracking results. In this direction, Kumar et al. [16] have integrated multiple features in the tracker appearance model in the PF framework. Rank-based adaptive fusion of features ensures the automatic boosting and suppression of the particles. Resampling technique based on meta-heuristic optimization, i.e. butterfly search optimization is exploited to address the PF shortcomings. Also, context-sensitive cue reliability is used for the tracker to be consistent with the environmental variations. Similarly, authors integrated multiple features using fuzzy-based fusion methodology along with the crow search optimization-based resampling technique [17]. Markov-based outlier detection mechanism is used to identify the low-performing particles to prevent the tracker from the erroneous update. However, experimental analysis is performed on a limited number of videos, which is insufficient to highlight the overall performance in tedious tracking challenges. Xiao et al. [7] have proposed to extract the target's contextual information to prevent the model from relearning from the background. A two-step estimation process is followed to prevent the tracker's drift. Initially, particles are initialized on the detected target and then, resampled to obtain the target's updated precise localization. On the other hand, Jiang et al. [8] have generated the confidence map for each pixel of the target for multiple features. The probability is calculated for each confidence map to determine whether the pixel belongs to the target or background. The fused confidence map is used to estimate the final localization of the target. In ref. [10], the authors have discounted each particle for three features. The conflict among the particles is resolved using the proportional conflict resolution rule (PCR-6) to obtain the final fused weight. Cuckoo search optimization is used to diversify the particles in the search space. However, Cai-Xia and Xin-yan [11] utilize the traditional sequential importance resampling to diversify the high-likelihood particles in the search space. Adaptive sub-patch multi-feature fusion is used to achieve robust tracking results. To reduce the processing time of each frame by the updated model of the tracker, authors have proposed to reduce the number of particles by decreasing the image size after marking the target [13]. Genetic algorithm is used in the resampling phase to choose the particles with higher weights. However, a hybrid tracking framework using PF and MS has been proposed, to optimize the computation task of the tracker. This approach has reduced the number of particles and is effective to handle abrupt motion and rotational changes. In summary, multi-feature-based trackers are effective to handle environmental variations. Also, strategies have been proposed to reduce the effective number of particles to minimize the processing computational efforts.

Further, to expand the efficiency of the trackers under the PF framework, information from multiple sensors is integrated [18, 19]. In this direction, it has been proposed to integrate the HSV data from the vision sensor and audio data from multiple audio speakers [18]. The number of particles is reduced significantly by incorporating audio information with visual information. Trackers have shown efficient results during occlusion and out-of-view challenges. Xiao et al. [19] have integrated IR information with visual data. The visual tracking template and IR tracking template are integrated using the new fusion rule to obtain the target's precise state estimation. Improved sampling along with adaptive occlusion handling is used to obtain robust tracking results. It has been proposed to integrate multiple information from the vision sensor with the thermal profile of the target obtained by thermal sensors [20]. Particles are redistributed and conflict among them is resolved using the PCR-5 rule. The tracker has obtained efficient tracking results during varying lighting and weather conditions. Zhang et al. [21] have integrated target sparse information and saliency extraction in IR data to track small objects. Fewer particles are used to update the target subspace to obtain its position. In summary, integrating multi-sensor data has improved tracking performance to a great extent. However, extraction and processing of multi-sensor data remain challenging as they enhance the tracker's complexity.

7.2 PARTICLE FILTER FOR VISUAL TRACKING

PF is also known as bootstrap filter [3] and condensation filter [22] for nonlinear non-Gaussian state estimation processes for visual tracking. PF utilizes a Bayesian framework for the state estimation process. PF has two steps procedure to obtain the desired posterior density function (PDF). The details of the state estimation using PF are as follows.

7.2.1 State estimation using particle filter

PF determines to approximate the PDF $\wp(S_t|x_t)$ of the target state when its prior PDF $\wp(S_{t-1}|x_{t-1})$ is available. Here, S_t and S_{t-1} represent the state vector for N particles for the frame at time t and $t-1$ of the video sequence along with respective x_t and x_{t-1} as a set of measurements.

For visual tracking, the state vector for each particle considering the rotational σ_t and scaling component $\varnothing_t$ using the random walk model [23] is represented as $S_t = \left(X_t, X_t^{'}, Y_t, Y_t^{'}, \sigma_t, \varnothing_t\right)$. Here, X_t and Y_t are Cartesian coordinates for the center of BB along with velocities $X_t^{'}$ and $Y_t^{'}$ in respective directions. Particles are evolved using the state model given by Eq. (7.1).

$$S_t = f\left(S_{t-1}, \eta_{t-1}\right) \tag{7.1}$$

where η_{t-1} is zero mean white noise and $f:\left(\mathcal{R}^{p'} \in \mathcal{R}^{p'} \rightarrow \mathcal{R}^{q'}\right)$ is a multi-component transition function. PDF $\wp(S_t|x_t)$ of the target state S_t is obtained into two steps: prediction and update. Initially, target PDF is obtained by random placement of particles manually. The subsequent state is estimated as the prior PDF of the target using Eq. (7.2).

$$\wp(S_t \mid x_t) = \int \wp(S_t \mid S_{t-1})\wp(S_{t-1} \mid x_{t-1})dS_{t-1} \tag{7.2}$$

Where particles are evolved using the multicomponent state model to predict the particles PDF $\wp(S_t|S_{t-1})$. The particle state is updated using the Bayes rule given by Eq. (7.3).

$$\wp(S_t \mid x_t) = \frac{\wp(x_t \mid S_t)\wp(S_t \mid x_{t-1})}{\wp(x_t \mid x_{t-1})} \tag{7.3}$$

In the above equation, the normalized denominator is obtained using Eq. (7.4).

$$\wp(x_t \mid x_{t-1}) = \int \wp(x_t \mid S_t)\wp(S_t \mid x_{t-1})dS_t \tag{7.4}$$

where $\wp(x_t|S_t)$ is the particle state obtained for each feature and posterior PDF $\wp(S_t|x_t)$ is used to compute the target state estimation as the weighted mean of the N particles. Details of the benefits and limitations of PF for visual tracking are discussed as follows.

7.2.2 Benefits and limitations of particle filter for visual tracking

PF is a powerful framework for solving tracking problems under Bayesian models. It uses weighted random samples to estimate PDF for determining the target's state [3]. As a result, PF can provide efficient and accurate target state estimation during tough tracking challenges. PF can effectively minimize the number of sampling patches during tracking and hence reduce the computational processing of the tracker. In addition, it can handle multimodal distribution generated by background clutters. PF can also adapt and expand to the dynamic environment during tracking for precise state estimation. In summary, PF has superiority over other existing algorithms due to its simplicity and less complexity in the tracking domain [16, 17].

PF has great potential to address the tracking problem effectively. However, PF has some inherent drawbacks which limit its applicability in the tracking domain: sample impoverishment and sample degeneracy. Sample impoverishment is the defect in PF causing particles to accumulate in a small area. Hence, the search space diversity is lost and ultimately particles with similar weights will be obtained. Sample degeneracy is a problem resulting from all the particles' weight lessening, so contributing little toward state estimation. To improve the accuracy and effectiveness of PF in the tracking domain, these defects should be resolved. Many researchers have proposed many solutions to address these shortcomings. To address the sample impoverishment problem, resampling techniques have been proposed [24–26]. Resampling techniques such as sequential importance resampling [11], adaptive resampling [25], and dynamic resampling [26] are exploited to replace the particles with new particles with improved weights. Although resampling methods cater to the sample impoverishment to a great extent this will lead to sample degeneracy problem in which all the particles accumulate in the same space.

To address the sample degeneracy, nature-inspired meta-heuristics optimization algorithms are exploited by many researchers to propose a reliable appearance model under the PF framework [1, 5, 15, 16]. Recently, some popular meta-heuristics optimization such as modified grey wolf optimization [1], ant colony optimization [15], butterfly search algorithm [16], and crow search optimization [17], are utilized to address the drawback of PF. In [1], authors have applied modified grey wolf optimization before resampling to spread them into the high-likelihood region to prevent sample impoverishment and degeneracy issue. However, to reduce the number of resampling particles, an outlier detection mechanism has been used [16]. In their work, particles are resampled by butterfly search optimization to diversify them in the search space. Authors have replaced the resampling step using ant colony optimization to reduce the impact of PF drawbacks [15].

Particles are populated as ants, and the best particles are obtained iteratively for the target state estimation. To summarize, meta-heuristics optimization-based PF tracking algorithms are efficient enough to be considered for real-time application. Also, the limitations of PF are addressed by preventing particles into the low-likelihood region. The processing of PF trackers is enhanced by computing only those particles whose contribution toward state estimation is significant. The next section will detail the PF tracking framework.

7.3 FRAMEWORK AND PROCEDURE

In this section, we discuss the tracking algorithm under the PF framework proposed by Kumar et al. [16], who exploited three features: color, LBP, and PHOG. A detailed explanation of the extraction of each of these features is given in Chapter 3, where we have discussed the various handcrafted feature extraction methods. Figure 7.1 illustrates the detailed tracking framework.

Initially, the target is detected in the first frame of the video sequence using the background GMM subtraction [27]. After this, N particles are instantiated around the centroid of the detected target. These particles evolved through the state model defined in Section 7.2.1. Each predicted particle is discounted for three features namely, color, LBP, and PHOG individually. These features are integrated using adaptive fusion based on non-linear ranking to obtain the final weight for each particle. The fusion approach will boost the significant particles and suppress the low-performing particles to create decision boundary for the outlier detection process. Outlier detection classifies the particles into significant particles and the low-performing particles as outliers to address the drawbacks of the PF. Outlier particles are further subjected to a resampling technique based on nature-inspired optimization. This resampling technique ensures the outliers will diversify in the search space to contribute to the target's state estimation by its two variables: sensor mobility and switch probability. Sensor mobility support in localizing the particles and switch probability propagates these particles in a high-likelihood area. The final state estimation of the target is computed as the weighted sum of the important particles and the resampled outliers. In addition, context-sensitive cue reliability is determined for each feature to discount each particle to make the tracker adaptive to dynamic environmental variations. The reference dictionary is updated using selective replacement of important particles. These steps ensure the consistent and temporal update of the tracker at each time step with environmental variations. The next section will discuss the adaptive fusion model for the multi-feature appearance model.

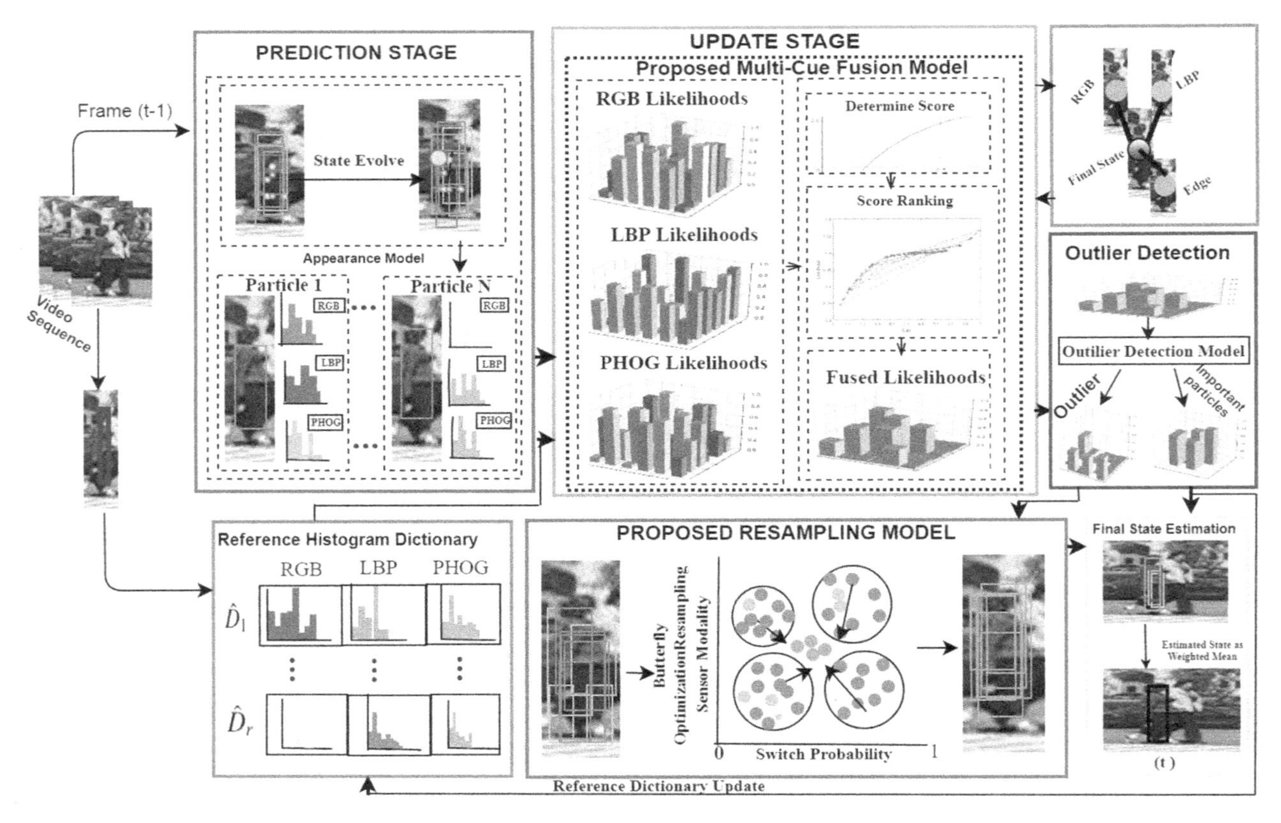

Figure 7.1 Architecture of the tracker under stochastic framework using PF. The prediction stage includes particle initialization and discounting each particle for multi-feature. The update stage deals with the adaptive integration of multi-feature likelihood. Outlier detection mechanism and resampling model address the drawback of the PF [16].

7.4 FUSION OF MULTI-FEATURES AND STATE ESTIMATION

In this section, we discuss the nonlinear ranking-based fusion of particles for each feature. This fusion approach ensures to boost of the significant particles and suppresses the low-performing particles for precise state estimation. Each n_{th} particle is discounted for three features assigned initial weights using Eq. (7.5).

$$w_f^n = \beta_f^n\left(B_{f,r}^n, F_f^n\right), n = 1,2,\ldots N \text{ and } f \in C,T,P \tag{7.5}$$

where N is the total number of particles, $\beta_f^n\left(B_{f,r}^n, F_f^n\right)$ represents the Bhattacharya's distance between the reference dictionary $B_{f,r}^n$, and the corresponding feature F_f^n and $f \in C, T, P$ for each feature namely, color, texture, and edge. The individual weight for each particle discounted for each feature is given by Eq. (7.6).

$$w_f^N = \begin{pmatrix} w_C^1 & w_C^2 \cdots & w_C^N \\ w_T^1 & w_T^2 \cdots & w_T^N \\ w_P^1 & w_P^2 \cdots & w_P^N \end{pmatrix} \tag{7.6}$$

Context reliability is calculated for each feature at $t - 1$ and used to determine reliability at time t as $R_{t,f} = (R_{t-1,C}, R_{t-1,T}, R_{t-1,P})$. The obtained reliability value for each feature is multiplied by the corresponding feature to obtain Eq. (7.7).

$$w_f^{N,R} = \begin{pmatrix} R_{t-1,C} \times w_C^1 & R_{t-1,C} \times w_C^2 \cdots & R_{t-1,C} \times w_C^N \\ R_{t-1,T} \times w_T^1 & R_{t-1,T} \times w_T^2 \cdots & R_{t-1,T} \times w_T^N \\ R_{t-1,P} \times w_P^1 & R_{t-1,P} \times w_P^2 \cdots & R_{t-1,P} \times w_P^N \end{pmatrix} \tag{7.7}$$

Next, the obtained weights of the particles for each feature are ranked in descending order. Each particle is ranked as $\mathfrak{R} = \mathfrak{R}_N, \mathfrak{R}_{N-1}, \ldots \mathfrak{R}_1$. Here, in a row particle with the highest rank is represented as $\mathfrak{R}_N$, and so on. The updated weighted matrix after ranking the particles is given by Eq. (7.8).

$$W_f^N = \begin{pmatrix} \mathfrak{R}_{N,C}\left(w_C^{1,R}\right) & \mathfrak{R}_{N-1,C}\left(w_C^{2,R}\right)\cdots & \mathfrak{R}_{1,C}\left(w_C^{N,R}\right) \\ \mathfrak{R}_{N,T}\left(w_T^{1,R}\right) & \mathfrak{R}_{N-1,T}\left(w_T^{2,R}\right)\cdots & \mathfrak{R}_{1,T}\left(w_T^{N,R}\right) \\ \mathfrak{R}_{N,P}\left(w_P^{1,R}\right) & \mathfrak{R}_{N-1,P}\left(w_P^{2,R}\right)\cdots & \mathfrak{R}_{1,P}\left(w_P^{N,R}\right) \end{pmatrix} \tag{7.8}$$

The obtained particle weights are normalized using row-wise min–max normalization and normalized weights are obtained as Eq. (7.9).

$$\dot{W}_f^N = \begin{pmatrix} W_C^1 & W_C^2 \cdots & W_C^N \\ W_T^1 & W_T^2 \cdots & W_T^N \\ W_P^1 & W_P^2 \cdots & W_P^N \end{pmatrix} \tag{7.9}$$

The normalized scores are subjected to function to transform $\dot{W}_f^N \rightarrow W_{\text{fus}}^n$. W_{fus}^n represents the final fused score and the transform function for n_{th} particle is given by Eq. (7.10).

$$W_{\text{fus}}^n = \frac{\dot{W}_C^n}{1 + \dot{W}_C^n \times \dot{W}_T^n \times \dot{W}_P^n} + \frac{\dot{W}_T^n}{1 + \dot{W}_C^n \times \dot{W}_T^n \times \dot{W}_P^n} + \frac{\dot{W}_P^n}{1 + \dot{W}_C^n \times \dot{W}_T^n \times \dot{W}_P^n} \tag{7.10}$$

Finally, the fused weights are subjected to a nonlinear function to boost and suppress the concordant and discordant features using Eq. (7.11).

$$\dot{W}_{f,t}^n = \frac{e^{W_{\text{fus}}^n} - e^{-W_{\text{fus}}^n}}{e^{W_{\text{fus}}^n} - e^{-W_{\text{fus}}^n}} \tag{7.11}$$

The final weights $\dot{W}_{\text{fus}}^n$ are assigned to the n_{th} particles at time t. After this, these particles are subjected to an outlier detection mechanism.

7.4.1 Outlier detection mechanism

The outlier detection mechanism classifies the particles into significant particles $\widehat{I_x}$ and outliers $\widehat{L_y}$. Outliers are low-performing particles that are most affected and have negligible weight. These particles do not contribute much toward state estimation and processing them requires excessive computations. Particles are classified using Eqs. (7.12) and (7.13).

$$\widehat{I_x} = W_f^n, \text{where } \dot{W}_f^n < \tau \tag{7.12}$$

$$\widehat{L_y} = W_f^n, \text{where } \dot{W}_f^n < \tau \tag{7.13}$$

where τ is a predefined threshold, $|x \cup y| = N$, $\widehat{I_x} \epsilon \left[X_{t,x}, Y_{t,y}\right]$, and $\widehat{L_y} \epsilon \left[X_{t,x}, Y_{t,y}\right]$.The so-obtained outliers $\widehat{L_y}$ are further subjected to resampling to prevent sample degeneracy.

7.4.2 Optimum resampling approach

Particles classified as outliers are resampled using the meta-heuristic optimization based on the butterfly optimization algorithm [28]. This resampling technique will disperse the outliers in the high-likelihood area by using its two parameters, sensor modality and switch probability. Sensor modality controls the convergence speed to determine the local maxima and switch probability determines optimal weight by switching between the local search and global search. The optimization will initialize the butterfly population as outliers and update their global position in the search space using Eq. (7.14).

$$L_y = \begin{cases} \widehat{L_y} + \left(\text{rnd}^2 \times b' - \widehat{L_y}\right) \times \mu_p \text{ if rnd} \leq \text{pob} \\ \widehat{L_y} + \left(\text{rnd}^2 \times \widehat{L_j} - \widehat{L_k}\right) \times \mu_p \text{ otherwise} \end{cases} \tag{7.14}$$

where j, $k \in k$, rnd ϵ [0, 1] and $L_y \; \epsilon \; [X_{t,x}, Y_{t,y}]$. b' is the best global position of particles in the current iteration. μ_p is $\mu_p = m_o I^a$. I represents the current position of the particles. m_o & $a \; \epsilon$ [0, 1] are the search controlling parameters.

7.4.3 State estimation and reliability calculation

Final state G_t of the target at time t is estimated by taking the weighted mean of the significant particles and the resampled particles using Eq. (7.15).

$$G_t = \frac{\sum_x \dot{W}_x \times \widehat{I_x} + \dot{\sum_y} \dot{W}_y \dot{\times} L_y}{\sum_{n=1}^{N} W_n} \tag{7.15}$$

where $G_t \in [X_t, Y_t]$ and $\widehat{I_x}$, L_y are the particle state determined by significant particles and resampled particles, respectively. Next, context cue reliability is determined for each feature for the tracker's quick adaptation to environmental variations. For this, L2-norm distance is computed between the final estimated state and the state estimated by each feature individually. The L2-norm distance is computed using Eq. (7.16).

$$d_{t,f} = \left\| G_t - G_{t,f} \right\| = \sqrt{\left(X_t - X_{t,f}\right)^2 + \left(Y_t - Y_{t,f}\right)^2}, f \epsilon C, T, P \tag{7.16}$$

where (X_t, Y_t) is the centroid of the final estimated state, and $(X_{t,f}, Y_{t,f})$ is the centroid of the state computed by each feature individually. The feature reliability is computed using Eq. (7.17).

$$R_{t,f} = \frac{\tanh\left(-u\left(d_{t,f}\right)+h\right)}{2} + 0.5, f \epsilon C, T, P \tag{7.17}$$

where u and h are constant terms. The context reliability calculated at time $t - 1$ is used to discount the particles at time t. These reliability values ensure the adaptive fusion of the multi-features in the update model of the proposed tracker. Reference dictionary of each feature is also updated by selective replacement of the significant particles. This process is repeated iteratively for all the frames in the video sequences. If the target is lost, the tracker is reinitialized to re-detect the target.

7.5 EXPERIMENTAL VALIDATION OF THE PARTICLE FILTER-BASED TRACKER

The experimental analysis of the PF-based tracker under tedious environmental conditions: illumination variations (IV), deformation (DEF), fast motion (FM), motion blur (MB), scale variations (SV), out-of-view (OV), full occlusion (FOC), partial occlusion (POC), background clutter (BC), and rotational challenges (ROC) is performed on the video sequences from OTB [29] and VOT [30] data sets. Also, the experimental results are compared with modern equivalents using robust performance metrics namely, CLE, F-measure [31], AUC, DP, Precision plot and success plot [29]. The description and computation procedure of all of these performance metrics is given in Chapter 4. Table 7.2 shows the considered video sequence under the attributed challenge.

7.5.1 Attributed-based performance

In this section, the performance of PF-based trackers is analyzed and compared. For this, we have performed an exhaustive performance evaluation under the following challenges. Initially, $N = 49$ particles are instantiated

Table 7.2 Details of video sequence under considered the attributed challenge

SN	*Attributed challenge*	*Video sequence*
1.	IV & DEF	*CarDark, Crossing, Singer1, Basketball, Shaking*
2.	FM & MB	*CarScale, Pedestrian1, Jumping, Soccer1, Tiger*
3.	SV	*Dancer, MountainBike, CarScale, Human7, Walking*
4.	FOC or POC	*Jogging1, Jogging2, Subway, Walking2, Tiger*
5.	BC & LR	*Bolt2, Car2, Skating1, Walking2, Surfer*
6.	ROC	*Singer2, Dancer, Football, Soccer, Skating1*

on the target detected by GMM subtraction [27]. The results are compared with modern equivalents PF [3], PF-PSO [32], CSPF [17], and MCPF [33]. Figures 7.2 and 7.3 illustrate the precision plot and success plot from the challenging video sequences, respectively.

7.5.1.1 Illumination variation and deformation

The challenging video sequences are considered under this challenge. On average of the outcome, the tracker has achieved the second highest average DP of 0.894 and the highest average AUC score of 0.723 as illustrated by Figures 7.2(a) and 7.3(a), respectively.

7.5.1.2 Fast motion and motion blur

Under this challenge, the tracker has achieved superior performance in comparison to other trackers. It is due to the rotational component of the random walk model used for the distribution of particles. On average of the outcome, the tracker has achieved the highest DP score of 0.688 and the highest AUC score of 0.599 as illustrated by Figures 7.2(b) and 7.3(b), respectively.

7.5.1.3 Scale variations

The tracker's performance under SV has outperformed the other trackers. The scaling component in the tracker's state model has catered for scale variations efficiently. Figures 7.2(c) and 7.3(c), has shown that the tracker has achieved the second highest average DP score of 0.77 as well as AUC score of 0.622, respectively.

7.5.1.4 Partial occlusion or full occlusion

Tracker performance under POC or FOC is mainly addressed by the adaptive update of the trackers by the reference dictionary. In addition, the impacted particle weight is also improved by the optimum resampling technique. The tracker has achieved average DP scores of 0.869 and 0.734 and the same is depicted in Figures 7.2(d) and 7.3(d), respectively.

7.5.1.5 Background clutters and low resolution

Under BC and LR challenges, the utilized complementary features along with context-sensitive reliability have been attributed to the performance of the tracker. Figures 7.2(e) and 7.3(e) have illustrated the second highest average DP score of 0.46 and AUC score of 0.473, respectively.

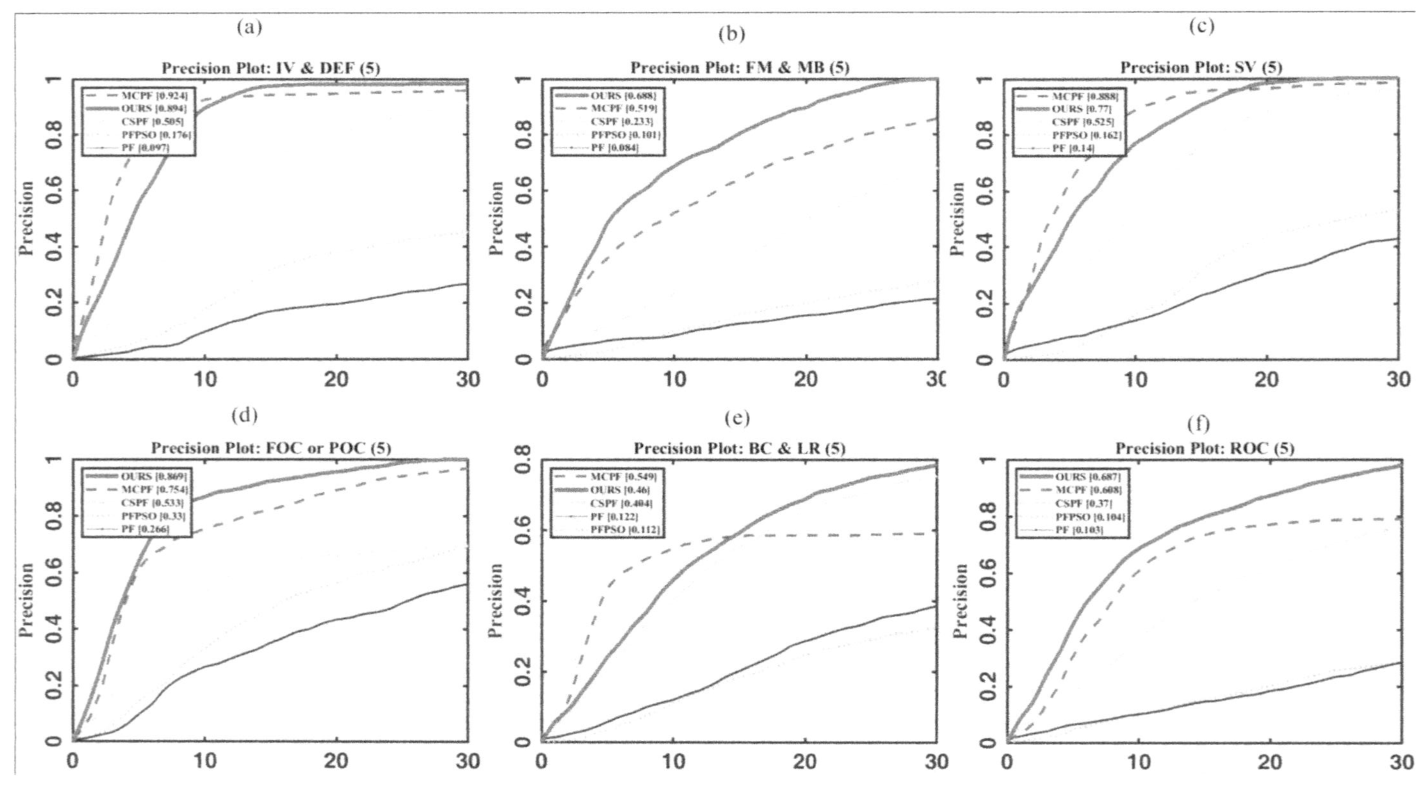

Figure 7.2 Precision plot under each attributed challenge. Average DP score is included in the brackets of legends.

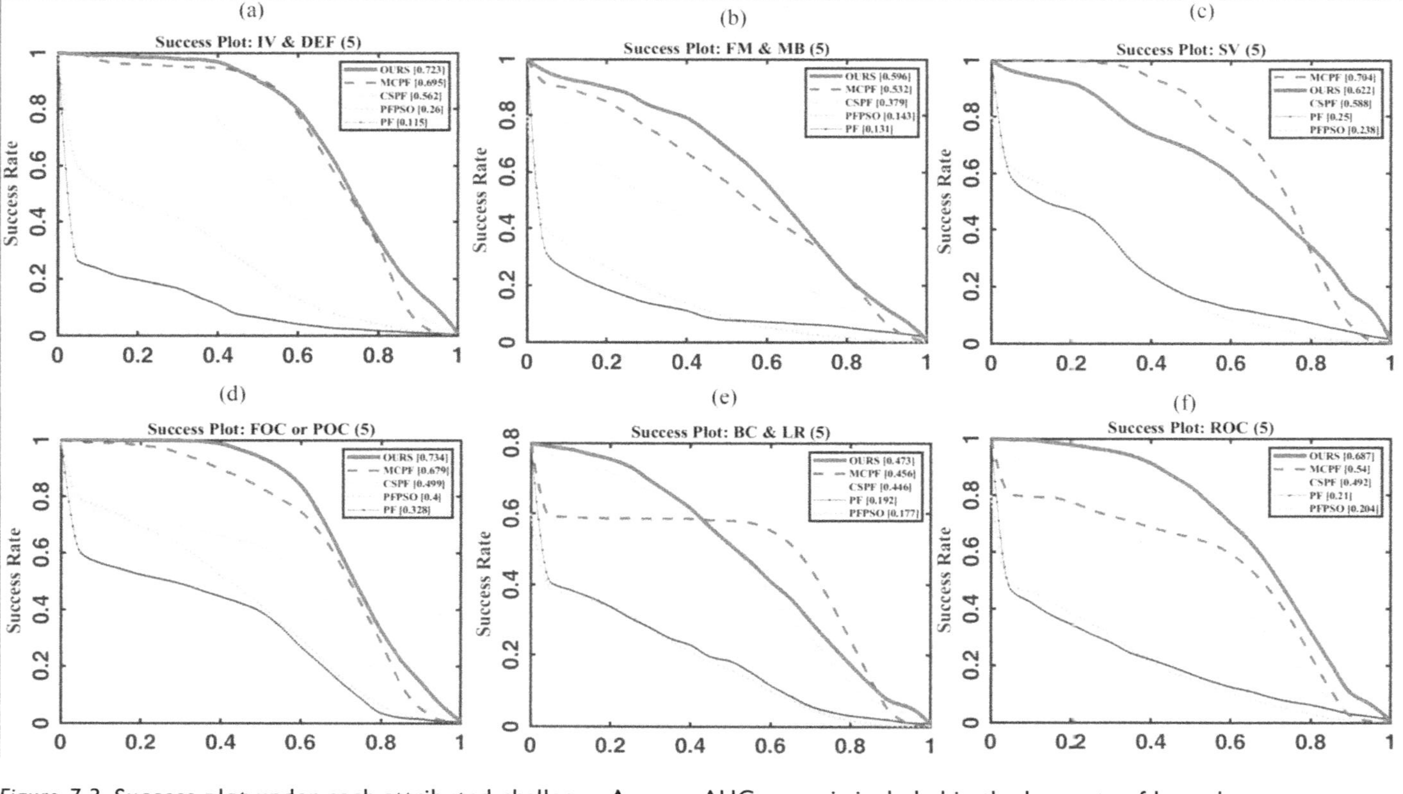

Figure 7.3 Success plot under each attributed challenge. Average AUC score is included in the brackets of legends.

7.5.1.6 Rotational variations

ROC challenge in the video sequences is addressed by the rotational and scaling component in the state model and the random walk model. In-plane and out-plane rotations are managed efficiently in comparison to other state-of-the-art. The highest average DP and AUC scores obtained were 0.467, and the same is depicted by Figures 7.2(f) and 7.3(f), respectively.

7.5.2 Overall performance evaluation

Tables 7.3 and 7.4 show the average CLE and average F-measure under variously attributed challenges. The proposed tracker has obtained an overall average CLE of 6.74 and an average F-measure of 0.747 on challenging video sequences. In addition, the overall precision plot and success plot on all the considered video sequences are illustrated in Figure 7.4. It has been observed that the tracker has obtained the overall highest average DP score and AUC as 0.743 and 0.644, respectively.

Table 7.3 Comparison of average CLE results obtained under each attributed challenge

SN	*Attributed challenge*	*PF*	*PF-PSO*	*MCPF*	*CSPF*	*OURS*
1.	IV & DEF	202.08	74.39	*5.23*	12.60	**6.01**
2.	FM & MB	124.57	59.32	*14.56*	25.69	**8.10**
3.	SV	70.03	28.88	**5.26**	10.66	*6.47*
4.	FOC or POC	68.86	31.26	*7.78*	24.41	**5.46**
5.	BC & LR	56.79	34.82	51.38	*9.54*	**8.30**
6.	ROC	103.01	64.47	43.69	*20.10*	**8.95**
7.	Overall	103.20	48.80	24.95	*16.45*	**6.74**

The first and second results are highlighted in **bold** and *italics*, respectively.

Table 7.4 Comparison of average F-measure results obtained under each attributed challenge

SN	*Attributed challenge*	*PF*	*PF-PSO*	*MCPF*	*CSPF*	*OURS*
1.	IV & DEF	0.136	0.332	*0.803*	0.692	**0.825**
2.	FM & MB	0.154	0.190	*0.641*	0.479	**0.708**
3.	SV	0.322	0.325	**0.814**	0.702	*0.725*
4.	FOC or POC	0.400	0.504	*0.791*	0.583	**0.838**
5.	BC & LR	0.239	0.228	0.508	*0.542*	**0.569**
6.	ROC	0.261	0.266	*0.626*	0.605	**0.796**
7.	Overall	0.253	0.387	*0.673*	0.592	**0.747**

First and second results are highlighted in **bold** and *italics*, respectively.

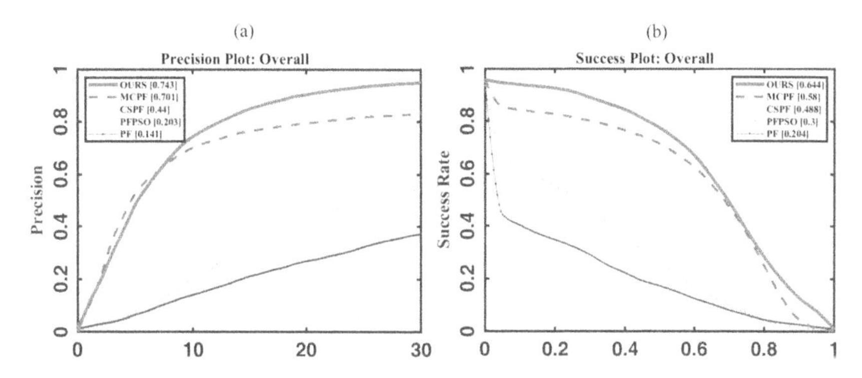

Figure 7.4 Overall performance plot (a) Precision (b) Success on challenging video sequences. Average DP and AUC scores are included in the brackets in the legend.

7.6 DISCUSSION ON PF-VARIANTS-BASED TRACKING

Visual tracking has been explored under various PF variants to improve its performance [15, 34–36]. PF can be integrated with other variants to reduce tracking failures during tedious environmental challenges [37–39]. Table 7.2 shows the representative work exploiting PF variants along with the description of salient features.

Under this, authors have proposed an iterative PF in which particles are generated decisively for better target estimation [40]. An incremental update scheme is used to update the object template adaptively during environmental variations. Trackers can address partial occlusion to prevent tracking failures. Similarly, Wang et al. [15] have used iterative PF to predict better particle states for efficient tracking. The so-generated particles are close to true posterior distribution and hence, can localize the target efficiently. Yoon et al. [41] have proposed fuzzy PF to address the noise variance during a complex environment.

Adaptive numbers of particles are adjusted during tracking to improve computational efficiency. The authors have fused the extended Kalman PF and least squares support vector regression to reduce the accuracy loss for target state estimation with fewer particles [35]. Extended Kalman PF is used to distribute the particles to obtain the posterior probability and least squares support vector regression is used to calibrate the motion trajectory of the target. However, MS and Kalman PF are integrated to overcome several tracking problems [42]. In this, authors have used MS as the master tracker primarily. To strengthen the tracker during occlusion, Kalman PF tracking is used to improve tracking results. Similarly, authors have integrated MS and Kalman PF but in a multi-feature appearance model to improve tracking efficiency [34]. Both algorithms cater to the limitations of each other when the target is present in the scene. Zhou et al. [36] have exploited hybrid PF based on the local and global information of the target.

Table 7.5 Description of representative work using PF-variants under the stochastic framework

SN	References	Year	Algorithm	Features	Summary
1.	Ho et al. [40]	2012	Iterative PF	PCA	Adaptive incremental model based on subspace learning.
2.	Yoon et al. [41]	2013	Fuzzy PF	HSV and edge	Fuzzy PF for improving the expected estimated states in the observation model.
3.	Zhou et al. [35]	2016	Extended Kalman PF	Color and motion	Utilized extended Kalman filter for particle distribution for better state estimation.
4.	Iswanto and Li [42]	2017	Kalman PF	HSV color	Utilize Kalman PF to address occlusion during tracking.
5.	Zhou et al. [36]	2017	Hybrid PF	Color and texture	Extracts target's local information via sparse and integrates with global information for precise tracking.
6.	Firouznia et al. [38]	2018	Chaotic PF	Color and motion	Localize target based on global information integrated with data from the local region.
7.	Qian et al. [43]	2018	Adaptive PF	Deep features	Utilize velocity and acceleration in dynamic model to optimize particle distribution.
8.	Iswanto et al. [34]	2019	Kalman PF	HSV color and color-texture	Mean shift-based tracking for handling occlusion and tedious environmental conditions.
9.	Wang et al. [15]	2020	Iterative PF	Locality sensitive histogram	Generate posterior distribution of particles using iterative PF.
10.	Dai et al. [37]	2020	Gaussian PF	CNN features, HOG	Extract features to represent the target location and scale.
11.	Nenavath et al. [39]	2022	Trigonometric PF	–	Improved PF performance by reducing the effective number of particles.
12.	Mozhdehi and Medeiros [44]	2022	Iterative PF	Convolutional features	Utilize prior distribution along with likelihood calculation using K-mean clustering at each frame.

Local information is extracted by using the multi-block sparse coding on the target. This information is combined with global tracking data to localize the target. Similarly, Firouznia et al. [38] have used chaotic PF in the local region to estimate the global target location. This method has utilized fewer numbers of particles in the search space to improve the tracker's efficiency for real-time tracking. In summary, the PF variant reduces the effective number of particles for the accurate state estimation of the target. Hence, the computational complexity of the tracker is reduced and improved results are obtained during complex environments in real-time videos.

In another line of work, PF variants are also explored in DL-based trackers [37, 43, 44]. For this, authors have proposed adaptive PF in a pre-trained CNN [43]. Particle distribution is optimized by considering the acceleration and velocity in the dynamic model. Numbers of particles are minimized by integrating the handcrafted features with the deep features in the appearance model. Motion information is used for reshaping and weighting the particles. Dai et al. [37] have localized targets by integrating a CF with Gaussian PF. Several weak classifiers are constructed along with reliability using an ensemble strategy to localize and scale the target as the weighted sum of each classifier. Gaussian PF has eliminated the requirement of resampling technique to prevent PF drawback and hence, computationally efficient in comparison to other PF-based trackers. Similarly, the authors have removed the requirement of particle resampling at each frame by integrating an iterative PF with CNN and CF [44]. Hierarchical convolutional response maps are generated for the particles generated around the predicted target. Particle likelihood is updated iteratively to refine the target location. However, the computation of CNN features has impacted the processing of the tracker and made it slow. In summary, deep features extracted from CNN are efficient in performance. These features are integrated into PF not only to improve its efficiency, but also to eliminate the need for resampling techniques. But processing the CNN feature is computationally expensive [45]. Interdependence between the particles can be explored for parallel processing to improve the tracker's speed.

7.7 SUMMARY

In this chapter, we have discussed the potential work under the PF framework. Traditional algorithms in which features are extracted from vision sensors as well as specialized sensors have been elaborated. Salient features are highlighted to identify the pros and cons of the existing work. PF drawbacks include sample impoverishment and sample degeneracy, as discussed. The existing work exploiting traditional resampling techniques and metaheuristic optimization-based resampling methods has been discussed to overcome PF drawbacks.

Furthermore, we have also discussed a PF-based multi-feature tracker based on nonlinear rank-based score fusion. Context-sensitive reliability is computed at each frame to improve the contribution of significant performing features during environmental variations. To address the shortcomings of PF, an optimum resampling technique based on butterfly search optimization has been proposed. To reduce the effect of sample impoverishment and improve computational efficiency, an outlier detection mechanism is exploited. This detects the affected particles so that their weight can be improved by the optimum resampling technique for precise state estimation. Experimental results are also obtained and compared with other state-of-the-art to prove the robustness of the tracker during complex environmental variations.

Apart from traditional PF, there exist other PF variants which not only have superior performance, but also eliminate the requirement of resampling in PF-based trackers. In other directions, deep features are also integrated with the PF framework to improve its efficiency during complex tracking variations. However, deep feature extraction utilized most of the processing power in the PF-based trackers. In order to improve computational efficiency, deep features-based PF trackers have not only eliminated the requirement of resampling technique but also reduces the effective number of particles for accurate and efficient tracking.

REFERENCES

1. Narayana, M., H. Nenavath, S. Chavan, and L.K. Rao, Intelligent visual object tracking with particle filter based on Modified Grey Wolf Optimizer. *Optik*, 2019. **193**: p. 162913.
2. Xue, X. and Y. Li, Robust particle tracking via spatio-temporal context learning and multi-task joint local sparse representation. *Multimedia Tools and Applications*, 2019. **78**(15): pp. 21187–21204.
3. Gordon, N.J., D.J. Salmond, and A.F. Smith. Novel approach to nonlinear/non-Gaussian Bayesian state estimation. in *IEE Proceedings F (Radar and Signal Processing)*. 1993. IET.
4. Rohilla, R., V. Sikri, and R. Kapoor, Spider monkey optimisation assisted particle filter for robust object tracking. *IET Computer Vision*, 2017. **11**(3): pp. 207–219.
5. Walia, G.S. and R. Kapoor, Intelligent video target tracking using an evolutionary particle filter based upon improved cuckoo search. *Expert Systems with Applications*, 2014. **41**(14): pp. 6315–6326.
6. Gao, M.-L., L.-L. Li, X.-M. Sun, L.-J. Yin, H.-T. Li, and D.-S. Luo, Firefly algorithm (FA) based particle filter method for visual tracking. *Optik*, 2015. **126**(18): pp. 1705–1711.
7. Xiao, J., R. Stolkin, M. Oussalah, and A. Leonardis, Continuously adaptive data fusion and model relearning for particle filter tracking with multiple features. *IEEE Sensors Journal*, 2016. **16**(8): pp. 2639–2649.
8. Jiang, H., J. Li, D. Wang, and H. Lu, Multi-feature tracking via adaptive weights. *Neurocomputing*, 2016. **207**: pp. 189–201.

9. Truong, M.T.N., M. Pak, and S. Kim, Single object tracking using particle filter framework and saliency-based weighted color histogram. *Multimedia Tools and Applications*, 2018. 77(22): pp. 30067–30088.
10. Walia, G.S. and R. Kapoor, Online object tracking via novel adaptive multicue based particle filter framework for video surveillance. *International Journal on Artificial Intelligence Tools*, 2018. **27**(06): p. 1850023.
11. Cai-Xia, M. and Z. Xin-Yan, Object tracking method based on particle filter of adaptive patches combined with multi-features fusion. *Multimedia Tools and Applications*, 2019. **78**(7): pp. 8799–8811.
12. Zhang, Z., C. Huang, D. Ding, S. Tang, B. Han, and H. Huang, Hummingbirds optimization algorithm-based particle filter for maneuvering target tracking. *Nonlinear Dynamics*, 2019. **97**(2): pp. 1227–1243.
13. Moghaddasi, S.S. and N. Faraji, A hybrid algorithm-based on particle filter and genetic algorithm for target tracking. *Expert Systems with Applications*, 2020. **147**: p. 113188.
14. Dash, P.P. and D. Patra, An efficient hybrid framework for visual tracking using Exponential Quantum Particle Filter and Mean Shift optimization. *Multimedia Tools and Applications*, 2020. **79**(29): pp. 21513–21537.
15. Wang, F., Y. Wang, J. He, F. Sun, X. Li, and J. Zhang, Visual object tracking via iterative ant particle filtering. *IET Image Processing*, 2020. **14**(8): pp. 1636–1644.
16. Kumar, A., G.S. Walia, and K. Sharma, Real-time visual tracking via multi-cue based adaptive particle filter framework. *Multimedia Tools and Applications*, 2020. **79**(29): pp. 20639–20663.
17. Walia, G.S., A. Kumar, A. Saxena, K. Sharma, and K. Singh, Robust object tracking with crow search optimized multi-cue particle filter. *Pattern Analysis and Applications*, 2020. **23**(3): pp. 1439–1455.
18. Kılıç, V., M. Barnard, W. Wang, and J. Kittler, Audio assisted robust visual tracking with adaptive particle filtering. *IEEE Transactions on Multimedia*, 2014. **17**(2): pp. 186–200.
19. Xiao, G., X. Yun, and J. Wu, A new tracking approach for visible and infrared sequences based on tracking-before-fusion. *International Journal of Dynamics and Control*, 2016. **4**(1): pp. 40–51.
20. Walia, G.S. and R. Kapoor, Robust object tracking based upon adaptive multicue integration for video surveillance. *Multimedia Tools and Applications*, 2016. 75(23): pp. 15821–15847.
21. Zhang, X., K. Ren, M. Wan, G. Gu, and Q. Chen, Infrared small target tracking based on sample constrained particle filtering and sparse representation. *Infrared Physics & Technology*, 2017. **87**: pp. 72–82.
22. Kumar, A., R. Jain, V. A. Devi, & A. Nayyar, (Eds.). *Object Tracking Technology: Trends, Challenges, Impact, and Applications*, 2023. Springer.
23. Brasnett, P., L. Mihaylova, D. Bull, and N. Canagarajah, Sequential Monte Carlo tracking by fusing multiple cues in video sequences. *Image and Vision Computing*, 2007. **25**(8): pp. 1217–1227.
24. Bolić, M., P.M. Djurić, and S. Hong, Resampling algorithms for particle filters: A computational complexity perspective. *EURASIP Journal on Advances in Signal Processing*, 2004. **2004**(15): pp. 1–11.
25. Wang, Z., Z. Liu, W. Liu, and Y. Kong. Particle filter algorithm-based on adaptive resampling strategy. in *Proceedings of 2011 International Conference on*

Electronic & Mechanical Engineering and Information Technology. 2011. IEEE.
26. Zuo, J., Dynamic resampling for alleviating sample impoverishment of particle filter. *IET Radar, Sonar & Navigation*, 2013. 7(9): pp. 968–977.
27. Stauffer, C. and W.E.L. Grimson. Adaptive background mixture models for real-time tracking. in *Proceedings. 1999 IEEE Computer Society Conference on Computer Vision and Pattern Recognition (Cat. No PR00149)*. 1999. IEEE.
28. Arora, S. and S. Singh, Butterfly optimization algorithm: A novel approach for global optimization. *Soft Computing*, 2019. **23**(3): pp. 715–734.
29. Wu, Y., J. Lim, and M.-H. Yang. Online object tracking: A benchmark. in *Proceedings of the IEEE Conference on Computer Vision and Pattern Recognition*. 2013.
30. Kristan, M., A. Leonardis, J. Matas, M. Felsberg, R. Pflugfelder, L. Čehovin Zajc, ... A. Eldesokey. The sixth visual object tracking VOT2018 challenge results. in *Proceedings of the European Conference on Computer Vision (ECCV) Workshops*. 2018.
31. Lazarevic-McManus, N., J. Renno, D. Makris, and G.A. Jones, An object-based comparative methodology for motion detection based on the F-measure. *Computer Vision and Image Understanding*, 2008. **111**(1): pp. 74–85.
32. Zhang, M., M. Xin, and J. Yang, Adaptive multi-cue-based particle swarm optimization guided particle filter tracking in infrared videos. *Neurocomputing*, 2013. **122**: pp. 163–171.
33. Zhang, T., C. Xu, and M.-H. Yang, Learning multi-task correlation particle filters for visual tracking. *IEEE Transactions on Pattern Analysis and Machine Intelligence*, 2018. **41**(2): pp. 365–378.
34. Iswanto, I.A., T.W. Choa, and B. Li, Object tracking based on mean shift and particle-Kalman filter algorithm with multi features. *Procedia Computer Science*, 2019. **157**: pp. 521–529.
35. Zhou, Z., D. Wu, and Z. Zhu, Object tracking based on Kalman particle filter with LSSVR. *Optik*, 2016. **127**(2): pp. 613–619.
36. Zhou, Z., M. Zhou, and J. Li, Object tracking method based on hybrid particle filter and sparse representation. *Multimedia Tools and Applications*, 2017. **76**(2): pp. 2979–2993.
37. Dai, M., G. Xiao, S. Cheng, D. Wang, and X. He, Structural correlation filters combined with a Gaussian particle filter for hierarchical visual tracking. *Neurocomputing*, 2020. **398**: pp. 235–246.
38. Firouznia, M., K. Faez, H. Amindavar, and J.A. Koupaei, Chaotic particle filter for visual object tracking. *Journal of Visual Communication and Image Representation*, 2018. **53**: pp. 1–12.
39. Nenavath, H., K. Ashwini, R.K. Jatoth, and S. Mirjalili, Intelligent trigonometric particle filter for visual tracking. *ISA Transactions*, 2022. **128**: pp. 460–476.
40. Ho, M.-C., C.-C. Chiang, and Y.-Y. Su, Object tracking by exploiting adaptive region-wise linear subspace representations and adaptive templates in an iterative particle filter. *Pattern Recognition Letters*, 2012. **33**(5): pp. 500–512.
41. Yoon, C., M. Cheon, and M. Park, Object tracking from image sequences using adaptive models in fuzzy particle filter. *Information Sciences*, 2013. **253**: pp. 74–99.
42. Iswanto, I.A. and B. Li, Visual object tracking based on mean shift and particle-Kalman filter. *Procedia Computer Science*, 2017. **116**: pp. 587–595.

43. Qian, X., L. Han, Y. Wang, and M. Ding, Deep learning assisted robust visual tracking with adaptive particle filtering. *Signal Processing: Image Communication*, 2018. **60**: pp. 183–192.
44. Mozhdehi, R.J. and H. Medeiros, Deep convolutional correlation iterative particle filter for visual tracking. *Computer Vision and Image Understanding*, 2022. **222**: p. 103479.
45. Kumar, A., G.S. Walia, & Sharma, K. (2020). Recent trends in multicue based visual tracking: A review. *Expert Systems with Applications*, **162**: p. 113711.

Chapter 8

Multi-stage and collaborative tracking model

8.1 INTRODUCTION

Visual tracking frameworks utilizing the various frameworks in the appearance model are categorized either as multi-stage methods or collaborative methods. Multi-stage methods focus on reduction of the computational load on the tracking model by incorporating a two-stage process, i.e. coarse-to-fine state estimation [1, 2]. Initially, the target position is roughly estimated, and then precisely localized in the final stage of estimation. Sometimes, unimportant background information is integrated with the target's appearance model. The processing of unimportant information with the target's relevant information increases the unnecessary computational load on trackers and makes the tracker unstable. To eliminate irrelevant information, two-stage tracking models are proposed [3, 4]. Hence, multi-stage trackers are not only computationally efficient, but also robust against various tracking challenges.

In another line of research, the robust tracker's appearance model is designed by exploiting the benefits of two tracking approaches: the generative and discriminative [5, 6]. Generative-based algorithms are efficient to handle illumination and scale variations, but not very effective at addressing occlusion and background clutters. On the other hand, discriminative-based algorithms are robust during complex occlusion and background clutter constraints. Hence, trackers based on a collaborative approach in their appearance model are popular nowadays. Also, the results from two independent trackers are collaborated to obtain rigorous tracking performance during complex environmental variations [7]. These trackers are adaptive to changing scenarios in long-term video sequences.

In this chapter, we have highlighted the representative work in the domain of multi-stage trackers and collaborative appearance model-based trackers. The algorithms are critically reviewed, and salient features are tabulated to determine the potential findings. In addition, we have discussed the multi-stage tracking frameworks. The multi-feature extraction methodology and fusion strategy are discussed in detail. The experimental results are determined and compared in line with the state-of-the-art to prove the superiority of the method.

DOI: 10.1201/9781003456322-8

8.2 MULTI-STAGE TRACKING ALGORITHMS

In this section, we have highlighted the recent work in the domain of multi-stage tracking algorithms. Multi-stage tracking algorithms are categorized under two classes: conventional multi-stage tracking and deep learning-based multi-stage tracking. Table 8.1 highlights the salient features of the representative work under each category. The work is arranged in ascending order of the published year. The details of potential work under each category are followed in turn.

8.2.1 Conventional multi-stage tracking algorithms

Conventional multi-stage tracking algorithms roughly localize the target during the first stage, and precise localization occurs during the next stage of estimation. These multi-stage approaches provide stable tracking results during tracking challenges. Under this, the fragment-based approach [8–11], sparse representation [3, 12], and superpixel-based approach [13] are popular.

Generally, fragment-based multi-stage trackers have extracted the optical flow using the Horn–Schunk method of the localized target during the initial stage of rough localization [8–11]. During precise localization, multi-features are extracted in the tracker's appearance model and adaptively fused to obtain stable tracking results. In ref. [8], authors have extracted two features: LBP and HOG in the tracker's appearance model. Features are adaptively fused by calculating the Euclidean distance between the target's extracted features for samples and the samples from the reference dictionary. K-means-based classifier is used to discriminate between the positive and the negative samples of the targets to prevent the processing of irrelevant information. The classifier not only prevents the erroneous update of the tracker but also enhances its computational efficiency. However, authors have extracted color and HOG in the tracker's appearance model [9]. The reliability-based adaptive update and fuzzy inference rule-based feature fusion is proposed to improve the tracking performance during complex environmental variations. On the other hand, Walia et al. [10] have integrated three features namely, intensity, LBP, and HOG using modified cross-diffusion based feature fusion strategy to obtain the unified feature. Unified graph feature fusion not only suppresses the affected features, but also enhances relevant features to minimize the tracker's computational efforts. Similarly, authors have extracted multi-feature during target precise localization [11]. These features are integrated using a nonlinear discriminating fusion strategy. However, there is no adaptive update of the trackers to prevent their drift during tedious tracking variations.

Target's coarse-to-fine state estimation using partial and structural information in sparse-based framework is proposed by [3]. An occlusion handling strategy is used to determine the occluded pixel along with

Table 8.1 Description of representative work under multi-stage tracking framework

SN	*References*	*Year*	*Algorithm*	*Update strategy*	*Summary*
1.	Jia et al. [3]	2016	Sparse representation	Incremental subspace learning	Template update along with occlusion detection to determine foreground pixels.
2.	Walia et al. [11]	2017	Fragment-based approach	Weighted mean-based reliability update	Nonlinear discriminating fusion of multi-feature in the appearance model.
3.	Zhang et al. [14]	2018	Reinforcement learning	Reward-based update	End-to-end training reinforcement learning architecture.
4.	Li et al. [15]	2018	Collaborative convolution operators	Gradient descent-based update	High-level feature extraction for capturing the semantic information efficiently.
5.	Xu et al. [12]	2019	Sparse discriminative CF	Incremental update	Obtain discriminating features by using specific temporal and spatial channel configurations.
6.	Kim et al. [13]	2019	Superpixels	Conditional statements based on selective update	Similarity and sampling measurement to select the superpixel for state estimation.
7.	Zhong et al. [16]	2019	Reinforcement learning	Stochastic gradient descent	KCF-based appearance model and supervision to prevent tracking failures.
8.	Walia et al. [10]	2019	Fragment-based approach	Reliability-based selective update	Cross-diffusion-based fusion of multi-features in the appearance model.
9.	Li et al. [17]	2019	CF	Stochastic gradient descent	Integrate both low-resolution and high-resolution features to address variations in target appearance.
10.	Wu et al. [18]	2020	PCA and background alignment	Incremental update	Utilize background alignment to determine the optimal occlusion mask to address partial occlusion.
11.	Kumar et al. [8]	2020	Target sampling-based approach	Reliability-based update	Euclidean distance-based fusion of multi-features to obtain unified robust features.
12.	Kumar et al. [9]	2020	Fragment-based approach	Reliability-based update	Fuzzy-based fusion of multi-features to obtain the robust feature invariant to environmental variations.
13.	Zgaren et al. [19]	2020	CNN and CF	Update control mechanism	Update the model using a control mechanism to prevent overfitting of the model.
14.	Han et al. [20]	2021	Siamese anchor-free proposal network	—	Multi-layer feature fusion to enhance discriminative capability during similar objects.
15.	Wen et al. [4]	2021	Attention mechanism	Linear update strategy	Quality feature map and precondition-based feature fusion for efficient tracking.
16.	Zeng et al. [21]	2022	Siamese point regression	—	Adaptive points on the target are converted to a BB to localize the target in the scene.

Note: The work is arranged in the ascending order of their published year.

likelihood-based consistent update of the tracker for efficient results during complex occlusion. To capture the target local information, an alignment pooling strategy is used for accurate target localization. In ref. [12], low-rank learning is proposed in sparse framework to eliminate the irrelevant features used by discriminative filters. Weighted coarse-to-fine strategy in discriminative CF is used to obtain efficient tracking results. Also, an optimization strategy based on augmented Lagrange multiplier is utilized to optimize tracking objective function.

The authors have explored the coarse-to-fine strategy by extracting the superpixels information associated with the target in a long-term tracking scenario [13]. During coarse estimation, the local search is performed to determine the target's superpixels and then similarity along with sampling is exploited for accurate and precise estimation of the target. The proposed method is not only computationally efficient, but also caters to the problem of local minima efficiently. In ref. [18], authors have utilized two-stage strategy to address the occlusion efficiently. Initially, the negative candidates are removed by the occlusion mask, and the observation model is reconstructed considering the previous frame to find the best optimal target location. PCA and background alignment strategy are integrated into the observation model to separate the target from the background during the occlusion challenge. In summary, conventional multi-stage models are computationally efficient as target localization is executed in various stages. The next section will discuss the deep learning-based multi-stage tracking models.

8.2.2 Deep learning-based multi-stage tracking algorithms

During the initial step, DL-based multi-stage tracking algorithms localize the target coarsely. After this, a series of estimations is performed to fine-tune the results. In this direction, reinforcement learning [14, 16], Siamese networks [20, 21], and CNN [19] based multi-stage tracking algorithms are proposed.

In ref. [14], the authors have utilized end-to-end deep networks for coarse-to-fine estimation of the target. The coarse estimation stage tracks the movement and scale variations of the target BB and the fine stage refines the BB boundaries for accurate estimation. There is a reward-based strategy to optimize the tracking results obtained by integrating the coarse and fine estimation. But Zhong et al. [16] have obtained motion information from deep recurrent reinforcement for coarse localization of the target. This information is exploited using kernelized CF for the target's accurate estimation during fine localization. Circulant matrices along with fast Fourier transform are utilized during fine estimation to determine the efficient candidate samples for accurate prediction of the target location. This prevents the tracker's degradation due to the update of the inaccurate candidate samples and hence, enhances tracking results.

The authors have proposed a three-stage tracking algorithm to obtain precise coordinates for target localization [20]. The three steps include feature extraction and fusion, classification and regression, and finally, validation and regression. To enhance the noise immunity and robustness of the tracker, multi-layer fusion is employed for constructing a target appearance model. Imbalance between the positive and negative candidate samples is removed using the target classification score and parameters are reduced for improving the tracker's processing. Zeng et al. [21] propose anchor-based tracking algorithms using end-to-end Siamese point regression. Initially, the points are positioned adaptively on the target to extract its location and shape. These points are fine-tuned for accurate localization by selecting the relevant hyper-parameters. The authors have utilized an attentional mechanism with transformer tracking for coarse-to-fine target estimation [4]. Coarse localization is achieved offline by segmenting the target. These target segments are fine-tuned online by regress BB. The quality target samples are obtained by the proposed swin transformer block to improve the tracking results.

CF is integrated with deep features in a multi-stage tracking strategy to enhance the tracker's discriminative power. In this direction, Zgaren et al. [19] have propose a two-stage procedure by integrating the benefits of CNN and CF. The target appearance variations are captured by deep features during coarse estimation and CF estimates the precise location of the target. The control update mechanism is adopted not only to learn new features, but also to prevent the tracker's erroneous update from the contaminated samples. However, the authors have integrated the results of the coarse tracker and refined tracker using cascade CF [17]. The coarse tracker captures the semantic features from the large search area whereas the refined tracker extracts the handcrafted features from the small search area based on the target's coarse localization. To summarize, DL-based multi-stage algorithms are robust against various tracking challenges. These trackers have superior discriminative power and high efficiency against various other tracking algorithms.

8.3 FRAMEWORK AND PROCEDURE

In this section, we will elaborate on the multi-stage tracking framework proposed in ref. [9], in which the authors have extracted multiple features in the target appearance model. A set of positive and negative samples is obtained by sampling the target, background, and the other target in the scene. A reference dictionary is initialized with one set of positive samples and two sets of negative samples. The motivation behind exploiting two sets of negative samples is to improve the discriminative power of the trackers from the background and in the presence of other similar objects in the scene.

Figure 8.1 illustrates the overview of the tracker's architecture. The tracker has adopted a two-stage strategy to enhance the efficiency of the

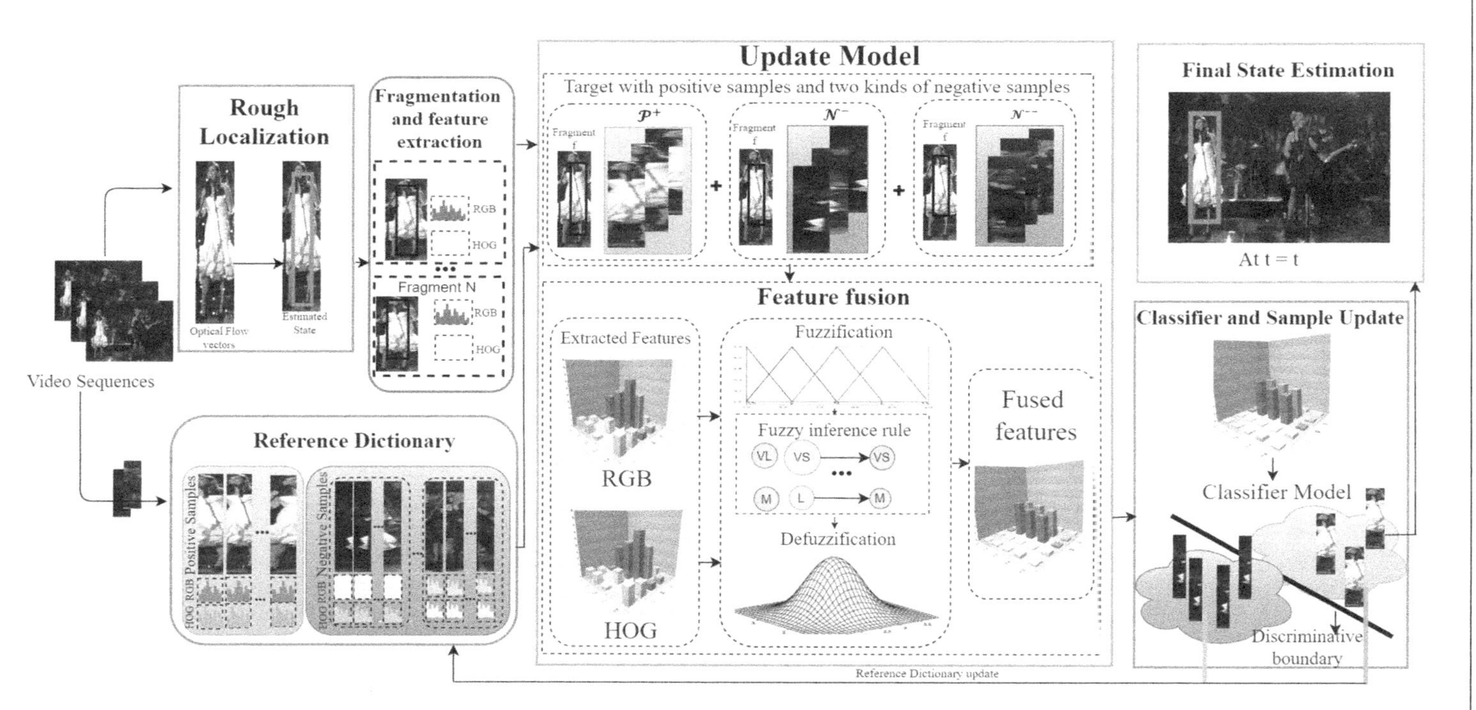

Figure 8.1 Overview of multi-stage tracker architecture. Final state estimation is a step procedure in which the target is roughly localized using optical flow and multi-features are extracted during precise localization [9].

proposed tracker during tracking variations. During rough localization, the target is localized using the optical flow algorithm proposed by Horn–Schunk [22]. Rough localized target is accurately estimated in the next stage of precise localization. Precise localization includes the fragmentation and multi-feature extraction step and adaptive multi-feature fusion to obtain unified features. The adaptive feature fusion strategy ensures the relevant features are captured and irrelevant features are suppressed in the final unified feature vector. Classifier and sample update stage classify the samples into positive and negative samples. This stage is responsible for the adaptive update of the tracker by selective replacement of the samples in the reference dictionary. Update stage ensures that the tracker is consistent with the changing environment and prevents the degradation of the tracker's performance with the contaminated samples. The final state is estimated with the weighted sum of the samples having high confidence score.

8.3.1 Feature extraction and fusion strategy

Initially, the target is roughly localized using the optical flow method. After this, multi-features are extracted for each candidate fragment: color and HOG. Details of the Horn–Schunk optical flow method, color, and HOG extraction are given in Chapter 3. During precise localization, the target is fragmented around the centroid obtained from the rough localization stage using a random walk model. Each fragment is discounted for multi-feature evaluation. Similarity between the target fragment and the samples stored in the reference dictionary is obtained by the Bhattacharya distance. The likelihood ($\zeta_{i,f}$) for each feature f in the ith fragment is obtained using Eq. (8.1).

$$\zeta_{i,f}\left(D, F_{i,f}\right) = \frac{1}{\sigma_f \sqrt{2\pi}} e^{\frac{\beta_{i,f}^2}{2\sigma_f^2}} \tag{8.1}$$

Where D is the reference dictionary consisting of one set of positive samples and two sets of negative samples. $F \in (C, H)$ is the feature set of extracted features: color and HOG. σ_f is the standard deviation calculated for each feature f and $\beta_{i,f}$ is the similarity computed using Bhattacharya distance for each feature f in the ith fragment and the corresponding sample in the reference dictionary.

8.3.1.1 *Multi-feature fusion and state estimation*

The extracted target features are adaptively fused using the fuzzy inference model. The fusion model ensures that the relevant details are captured correctly to obtain a robust unified feature. The fuzzy model parameters namely, fuzzy nearness and correlation coefficient are exploited in the fusion model [23]. Fuzzy nearness $M_n \in \mathcal{R}^{s \times s}$ computes the similarity between the ith fragments using Eq. (8.2).

$$M_{n,i}\left(C_j, \dot{H}_k\right) = \frac{\sum_{p=1}^{n} \min\left(\zeta_{C_j}^{p}, \zeta_{\dot{H}_k}^{p}\right)}{\sum_{p=1}^{n} \max\left(\zeta_{C_j}^{p}, \zeta_{\dot{H}_k}^{p}\right)}, j, k = 1, 2, \ldots, s \tag{8.2}$$

The correlation coefficient computes the degree of conflict between two sources of evidence. The correlation coefficient matrix $CC \in \mathcal{R}^{s \times s}$ computes the conflicts between two features for the ith fragments using Eq. (8.3).

$$CC\left(C_j, \dot{H}_k\right) = \begin{cases} \dfrac{\zeta_{C_j}^{\max} + \zeta_{\dot{H}_k}^{\max}}{2}, & \text{if } \zeta_{C_j}^{\max} = \zeta_{\dot{H}_k}^{\max} \\ \dfrac{\zeta_{C_j}^{\min} + \zeta_{\dot{H}_k}^{\min}}{2}, & \text{if } \zeta_{C_j}^{\max} \neq \zeta_{\dot{H}_k}^{\max} \end{cases} \tag{8.3}$$

The relationship between M_n and CC is represented using the fuzzy theory. These two parameters are fuzzified using the fuzzification process. For very small S^v, small S, medium M^s, large L and very large L^v values, sample set $\dot{S}$ is defined as $\dot{S} = \left\{ S^v, S, M^s, L, L^v \right\}, x \in \left[M_n \text{ or } CC\right]$. Membership functions are computed for each set using Eqs. (8.4)–(8.8).

$$\mu_{S^v}\left(x\right) = \frac{1}{1 + e^{-a_{S^v}(x - b_{S^v})}} \tag{8.4}$$

$$\mu_s\left(x\right) = e^{-(x - a_s)^2 / b_s^2} \tag{8.5}$$

$$\mu_{M^s}\left(x\right) = e^{-\left(x - a_{M^s}\right)^2 / b_{M^s}^2} \tag{8.6}$$

$$\mu_L\left(x\right) = e^{-(x - a_L)^2 / b_L^2} \tag{8.7}$$

$$\mu_{L^v}\left(x\right) = \frac{1}{1 + e^{-a_{L^v}\left(x - b_{L^v}\right)}} \tag{8.8}$$

Table 8.2 Fuzzy inference rules

$\dot{S}$		M_n				
		S^v	S	M^s	L	L^v
CC	S^v	S^v	S^v	S^v	M^s	L
	S	S^v	S^v	S	M^s	L
	M^s	S^v	S	M^s	M^s	L
	L	L	L	M^s	L^v	L^v
	L^v	L	L	L	L^v	L^v

Where the linguistic variables are defined for each membership function in the sample set as a_{S^v}, b_{S^v}, a_s, b_s, a_{M^s}, b_{M^s}, a_L, b_L, a_{L^v}, b_{L^v}. The relationship between M_n and CC is represented by fuzzy inference rules using Table 8.2.

The crisp values are obtained by the COG method for similarity matrix $\dot{\mathcal{L}}$. COG values between M_n and CC is computed using Eq. (8.9).

$$\text{COG}_{M_n,CC} = \frac{\sum_{M_n}\sum_{CC}\mu_{\dot{\mathcal{L}}}(x)x}{\sum_{M_n}\sum_{CC}\mu_{\dot{\mathcal{L}}}(x)} \tag{8.9}$$

These values are passed through a defuzzification process, and the similarity value for the ith fragments is computed using Eq. (8.10).

$$\mathcal{L}_i = \frac{\sum_{M_n}\sum_{CC}\chi_{M_n,CC}\times\text{COG}_{M_n,CC}}{\sum_{M_n}\sum_{CC}\chi_{M_n,CC}\left(\dot{\mathcal{L}}\right)} \tag{8.10}$$

Where $\chi_{M_n,\ CC}$ is the fuzzy control rule represented as $\chi_{M_n,CC} = \min[\mu_{M_n}\left(\dot{\mathcal{L}}_{M_n}, \mu_{CC}\left(\dot{\mathcal{L}}_{CC}\right)\right]$. The unified feature vector for the ith fragment is obtained by column-wise normalization of $\mathcal{L}_i$ using Eq. (8.11).

$$U(F_i) = \frac{\mathcal{L}_i(F_i)}{\sum_{i=1}^{s}\left|\mathcal{L}_i(F_i)\right|} \tag{8.11}$$

Next, the obtained unified feature is applied to a random forest classifier to classify the candidate fragments. For this, bagging algorithms are utilized which ensemble the several weak classifiers to generate strong classifiers. The classifier assigns the value [0,1] to each candidate fragment based on its resemblance to a related fragment learned during training. Also, the confidence score is generated for each fragment, and fragments with high scores are averaged to obtain the final score using Eq. (8.12).

$$\rho_i = \frac{e^{\iota}}{\sum_{j=1}^{I} e^{\iota,j}} \tag{8.12}$$

Where I is the number of high confidence score samples with score ι. The final target location at time t is determined using Eq. (8.13).

$$G_{t,(x,y)} = \sum_{i=1}^{I} \rho_i F_i(x,y) \tag{8.13}$$

Where $F_i(x, y)$ is the centroid for the ith fragment. In case the tracker lost the target, the tracker is reinitialized for the whole process to redetect the target again. The tracker is made adaptive to the tracking variations by selectively replacing the positive and negative samples from the reference dictionary.

8.3.2 Experimental validation

The performance of the proposed tracker is evaluated under tedious environmental conditions. For this, we have selected challenging video sequences from the two public data sets, namely OTB [24] and VOT [25]. The details of the attributed challenge and the considered video sequence are tabulated in Table 8.3.

For fair performance evaluation, the tracker's experimental results are compared with state-of-the-art equivalents: MEEM [26], STAPLE [27], DCFCA [28], and UGF [10]. Robust performance metrics namely, CLE, F-measure [29], AUC, DP, Precision plot, and success plot [24] are exploited

Table 8.3 Details of video sequence under considered the attributed challenges

SN	*Attributed challenge*	*Video sequence*
1.	IV & DEF	*CarDark, Crossing, Singer1, Basketball, Shaking*
2.	FM & MB	*CarScale, Pedestrian1, Jumping, Soccer1, Tiger*
3.	SV	*Dancer, MountainBike, CarScale, Human7, Walking*
4.	FOC or POC	*Jogging1, Jogging2, Subway, Walking2, Tiger*
5.	BC & LR	*Bolt2, Car2, Skating1, Walking2, Surfer*
6.	ROC	*Singer2, Dancer, Football, Soccer, Skating1*

for experimental evaluation. The description and computation procedure of all of these performance metrics is given in Chapter 4.

Initially, candidate fragments are generated on the target. The numbers of generated fragments are highly dependent on the target's size. Smaller targets are represented by fewer fragments in comparison to large-size targets with high resolution. The reference dictionary is instantiated with a set of 40 fragments for positive and negative samples each. The performance of the tracker under variously attributed challenges is elaborated as follows. Figures 8.2 and 8.3 illustrate the precision plot and success plot from the challenging video sequences, respectively.

8.3.2.1 Illumination variation and deformation

The challenging video sequences are considered under this challenge. The tracker achieved average DP of 0.809 and average AUC score of 0.715, as illustrated by Figures 8.2(a) and 8.3(a), respectively.

8.3.2.2 Fast motion and motion blur

Under this challenge, the tracker has achieved superior performance in comparison to other trackers. It is due to the consistent update of the tracker reference dictionary with the relevant fragments. The random forest classifier classifies the samples with high confidence scores to estimate the target position accurately. On average of the outcome, the tracker has achieved the highest DP score of 0.817 and the highest AUC score of 0.603 as illustrated by Figures 8.2(b) and 8.3(b), respectively.

8.3.2.3 Scale variations

The tracker's performance under SV is comparable to state-of-the-art. Tracker's performance under scale variations is degraded as the sample dictionary is contaminated by irrelevant fragments. Figures 8.2(c) and 8.3(c) show that the tracker has achieved average DP score of 0.575 and the second highest AUC score of 0.664, respectively.

8.3.2.4 Partial occlusion or full occlusion

Tracker performance under POC or FOC is attributed mainly to the incorporation of complementary features in the tracker's appearance model. In addition, the fuzzy-based fusion model ensures the adaptive update of the tracker in the presence of this challenge. The tracker has achieved the highest average DP score of 0.894 and 0.762 and the same is depicted in Figures 8.2(d) and 8.3(d).

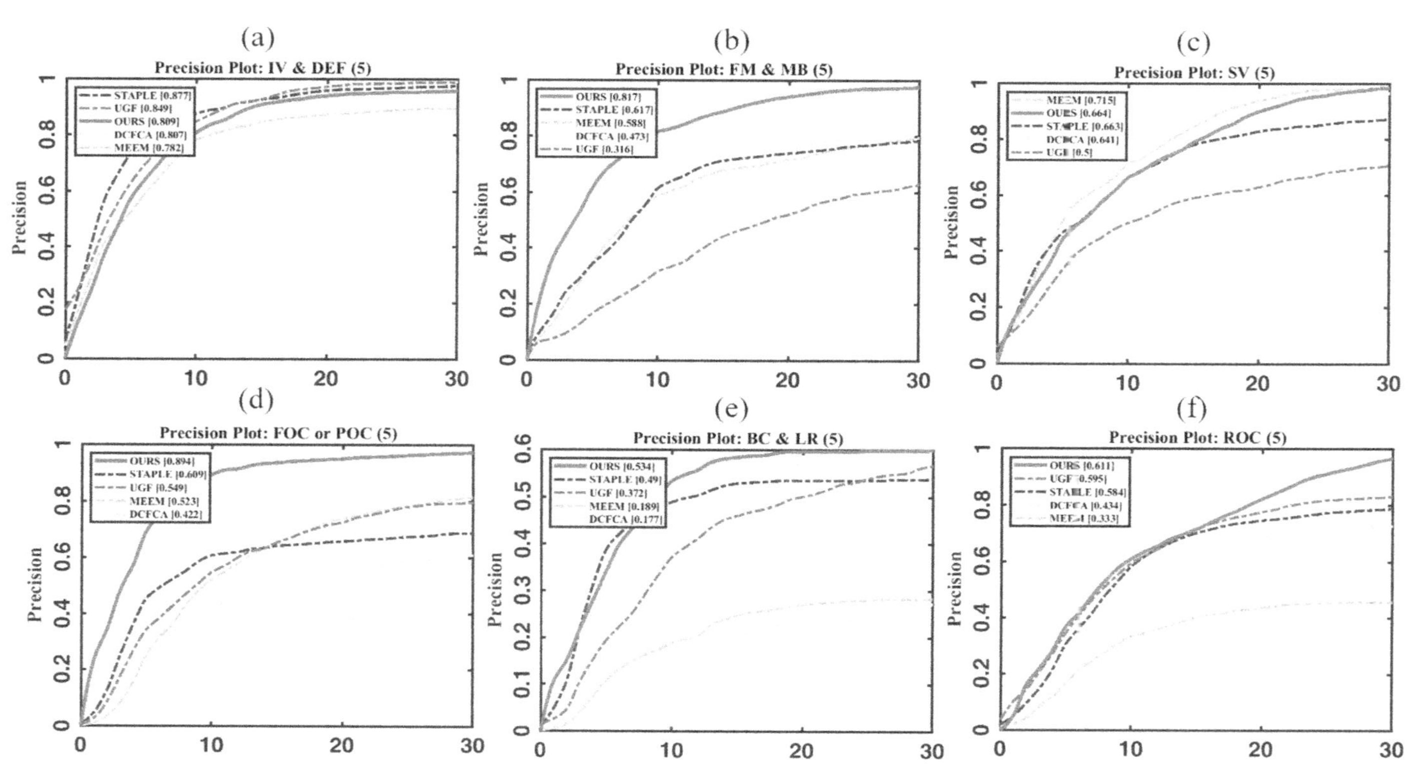

Figure 8.2 Precision plot under each attributed challenge. Average DP score is included in the brackets of legends.

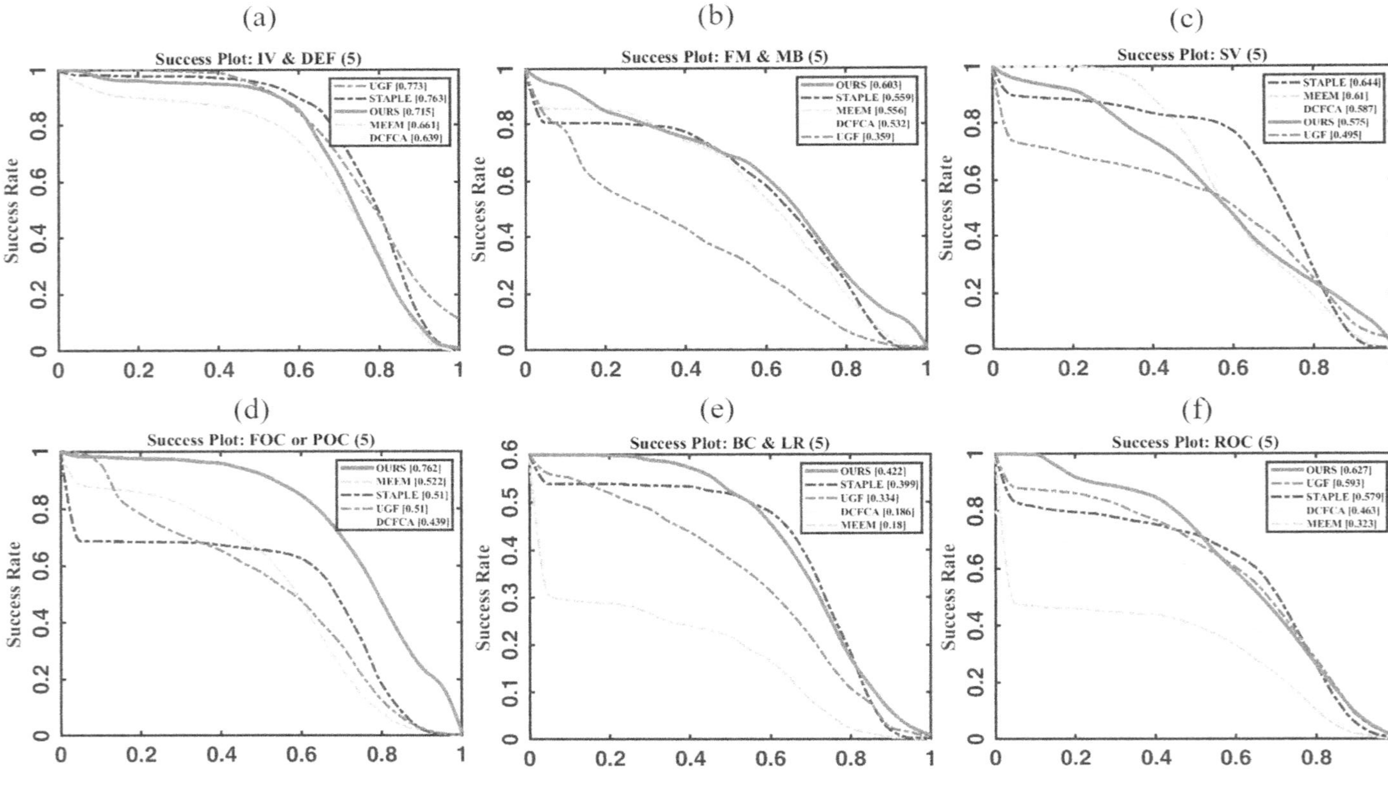

Figure 8.3 Success plot under each attributed challenge. Average AUC score is included in the brackets of legends.

8.3.2.5 Background clutter and low resolution

Under BC and LR challenges, the utilized complementary features have been attributed to the performance of the tracker. Also, the classifier can indicate positive and negative fragments efficiently and prevents dictionary erroneous updates. Figures 8.2(e) and 8.3(e) have illustrated the highest average DP score of 0.534 and AUC score of 0.422, respectively.

8.3.2.6 Rotational variations

ROC challenge in the video sequences is primarily addressed by the HOG feature exploited in the tracker's appearance model. Fuzzy-based fusion model ensures the unified feature is robust as HOG supersedes the color feature during ROC variations. Also, the rotational component in the random walk model manages the in-plane and out-of-plane rotations efficiently in comparison to other state-of-the-art equivalents. On average outcome, the highest average DP of 0.611 and highest AUC score of 0.627 are obtained for the tracker under this challenge and the same is depicted by Figures 8.2(f) and 8.3(f), respectively.

8.3.2.7 Overall performance comparison

Tables 8.4 and 8.5 show the average CLE and average F-measure under various attributed challenges. The proposed tracker has obtained an overall average CLE of 6.89 and an average F-measure of 0.746 on challenging video sequences. In addition, the overall precision plot and success plot on all the considered video sequences are illustrated in Figure 8.4. It has been observed that the tracker has obtained the overall highest average DP score and AUC as 0.755 and 0.646, respectively.

Table 8.4 Comparison of average CLE results obtained under each attributed challenge

SN	*Attributed challenge*	*MEEM*	*STAPLE*	*DCFCA*	*UGF*	*OURS*
1.	IV & DEF	11.58	6.84	*6.83*	**5.31**	7.12
2.	FM & MB	26.32	25.51	*19.97*	32.31	**6.81**
3.	SV	**7.57**	13.79	9.34	36.83	*8.64*
4.	FOC or POC	25.01	50.89	55.30	*17.55*	**5.86**
5.	BC & LR	64.75	8.95	68.12	*6.55*	**3.10**
6.	ROC	95.65	21.95	*10.74*	31.73	**10.36**
7.	Overall	41.02	22.21	32.88	*21.03*	**6.89**

Note: First and second results are highlighted in **bold** and *italics*, respectively.

Table 8.5 Comparison of average F-measure results obtained under each attributed challenge

SN	*Attributed challenge*	*MEEM*	*STAPLE*	*DCFCA*	*UGF*	*OURS*
1.	IV & DEF	0.760	*0.852*	0.753	**0.862**	0.817
2.	FM & MB	0.657	**0.738**	0.633	0.464	*0.706*
3.	SV	**0.743**	0.510	*0.708*	0.570	0.691
4.	FOC or POC	*0.639*	0.577	0.513	0.632	**0.848**
5.	BC & LR	0.214	*0.454*	0.227	0.401	**0.488**
6.	ROC	0.369	0.664	0.553	*0.689*	**0.742**
7.	Overall	0.579	*0.669*	0.564	0.647	**0.746**

Note: First and second results are highlighted in **bold** and *italics*, respectively.

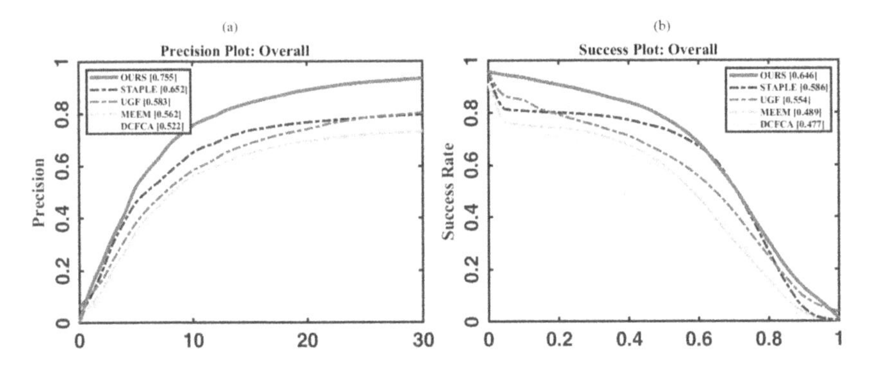

Figure 8.4 Overall performance plot (a) Precision (b) Success on challenging video sequences. Average DP and AUC scores are included in the brackets in the legend.

8.4 COLLABORATIVE TRACKING ALGORITHMS

For developing a strong appearance model invariant to environmental variations, generative and discriminative models have collaborated in the tracker's framework. Table 8.6 shows the salient features of the representative work developed using collaborative approach. In this direction, authors have exploited the generative approach using weighted structural local sparse in the tracker's appearance model and support vector machine-trained discriminative classifier [30]. The spatial configuration of the target local patch is weighted for developing the appearance model and the discriminative classifier is incrementally updated in a Bayesian framework. The tracker has shown efficient results in challenging scenarios. However, Dai et al. [31] have exploited the hash algorithm under modified PF to develop a generative appearance model. A discriminative classifier based on spatio-temporal context model is used for fine classification by maximizing the

Table 8.6 Description of representative work under collaborative tracking framework

SN	References	Year	Algorithm	Update strategy	Summary
1.	Duffner and Garcia [32]	2016	Contextual motion density and discriminative classifier	Probabilistic-based dynamic update	Specific background regions for negative samples enhance the discriminability of the tracker to extract the target from the foreground.
2.	Zhuang et al. [34]	2016	Block-based incremental learning and patch subspace learning	Block-based local PCA online update	Discriminative model addresses the target's appearance variations and the generative model handle occlusion.
3.	Tu et al. [35]	2017	ℓ_2 regularized least square and temporal information	Adaptive template update with multimemory	Cost function incorporated with smoothness term to improve the tracker's robustness.
4.	Zhou et al. [6]	2017	Coding method for object representation and graph regularized discriminant analysis	Adaptive online update	Final state estimation using nearest neighbor with an online adaptive update to address tracking variations.
5.	Dou et al. [30]	2017	Structural local sparse with patch matching	Incremental learning update	Positive and negative sample pooling with patch dictionary based on sparse learning.
6.	Liu et al. [36]	2018	Salient sparse generative and discriminative model	Adaptive template update	Salient feature map for the effective update of the template during occlusion.
7.	Wang et al. [37]	2018	Multi-task sparse with template matching	Adaptive template update	Confidence value-based discriminating strategy to separate target from the background.
8.	Liu et al. [7]	2018	Context tracker and kernelized CF tracker	Reliability-based adaptive update	Target model-based redetection strategy to again localize the lost target in the scene.
9.	Dai et al. [31]	2018	Hash algorithm based on spatio-temporal information	Weighted template update	Confidence map-based spatio-temporal extraction of target's feature.
10.	Yi et al. [2]	2018	Sparse representation and global discriminant layer	Dual threshold-based update	Spatial information along with template update strategy to deal with complex occlusion.
11.	Bao et al. [33]	2019	Patch-based strategy and contextual information	Adaptive local patch update	Confidence map strategy to handle complex occlusion during tracking.
12.	Kumar et al. [1]	2020	Global subspace and Bayesian inference framework	Incremental update model	Global and local robust coding with occlusion map generative to address tracking challenges.
13.	Zhao et al. [38]	2020	Ridge regression and multi-scale local sparse	Online interpolation-based update	Discriminative model update for precise localization and to handle target scale variations.
14.	Kong et al. [5]	2020	Multi-scale method and CF	Adaptive update	Patch-based tracking with weighted similarity calculation to handle complex occlusion situations.

Note: These works are arranged in order of published year.

confidence map. This step determines the more precise and accurate target location in challenging videos. Authors have exploited visual features along with motion prediction in a PF framework [32]. Online discriminative classifier based on motion contextual density function is used to quickly discriminate the target during distractions. The method can cope with changing variations and severe target deformation as both foreground and background information is updated dynamically. Contextual information in an adaptive PF environment has shown better results in comparison to other state-of-the-art. Also, contextual information is exploited with local patches for accurate tracking results [33]. Double BBs are used for object representation consisting of the target and the background for better feature extraction. Confidence maps are constructed to evaluate whether the local patches are affected due to occlusion or not. The contextual information and the target's patch likelihood are collaborated to provide robust and stable tracking results.

To improve the efficiency of the tracker, the authors have exploited graph regularized as discriminative approach and coding method-based generative approach [6]. The hybrid appearance model is developed by collaborating the likelihoods obtained from both approaches by using the nearest neighbor classifier. Target appearance is represented by global and local information to distinguish the target from the background. Wang et al. [37] have collaborated the subspace learning as generative approach and confidence value-based discriminative approach in the tracker's appearance model. Target's holistic information is exploited in the generative model based on PCA. The discriminative approach is based on multi-task sparse learning assigning dynamic weight to the samples based on their resemblance either to the target or background. ADMM-based optimization is used to localize the target accurately. But the authors have utilized multi-scale local sparse coding using PF in the generative model [38]. Two ridge regression models are included in the discriminative model of the tracker. Tracker is updated online consistently using linear interpolation to address the target's scale variations. Yi et al. [2] have proposed a hierarchical tracking framework consisting of three appearance models. Out of these, two are generative models based on local histogram and weighted alignment pooling. The third is a discriminative model based on sparse representation. The three models are fused based on the similarity and confidence score along with the online update of the dictionary. The generative model generates the target template by removing the background to better discriminate the target from the background in the discriminative model.

For quick redetection of a lost target in a scene, the authors have integrated the target's motion and context information [7]. Results from two trackers, target and context, are collaborated to prevent the tracker's drift during long-term tracking. This information is useful to discriminate targets during background clutter. However, Tu et al. [35] have integrated target motion information with holistic data. ℓ_2 regularized least square method and sparse coding is utilized in the tracking scheme to develop an efficient target appearance model. The relationship between the consecutive frames is modeled by introducing smooth terms in the regularizer. Motion and appearance information is collaborated in PF to build a robust tracker which can address challenging scenarios efficiently.

To deal with heavy occlusion scenarios, authors have collaborated with multi-scale methods along with discriminative CF [5]. Independent trackers based on the CN model with color features and kernelized CF with HOG features are integrated with CF to generate a weighted response. A classifier box based on similarity distances determines the occlusion and selects the appropriate tracker for accurate localization. In ref. [34], the authors have collaborated generative models based on subspace learning and discriminative model based on deep learning to model the target efficiently. Positive and negative patches are generated from the target and its background to fine-tune the tracker based on the offline update. Occlusion is addressed by the block-based PCA approach. Authors have collaborated on salient-based sparse discriminative and generative models for target localization [36]. A discriminative model is used to discriminate the target from the background using the salient feature map generated by the generative model. The salient feature maps are suitable to prevent the tracker's drift during occlusion and improve the effectiveness of the tracker. In summary, collaborative-based trackers are based on the mixture of two strategies that compensate for each other to deal with incorrect predictions during complex environmental conditions [39].

8.5 SUMMARY

In this chapter, we have summarized the various works in the domain of multi-stage tracking and collaborative tracking. Multi-stage trackers are computationally efficient as they estimate the target state in multiple stages. These trackers not only provide accurate tracking results but also reduce the processing of the tracker's observation model. In addition, we have also discussed a multi-stage tracking framework utilizing optical flow during rough localization and multi-features along with random forest classifier for the target's precise estimation. The tracker is consistently updated by selective replacement of the samples from its reference dictionary. Tracking results are compared with other potential state-of-the-art equivalents to highlight the stability of the multi-stage tracker during tough environmental conditions.

Collaborative trackers combine the generative model and discriminative model which address each other's shortcomings during target tracking. Typically, generative trackers are based on the target appearance model only. Hence, these trackers alone are not able to cope with the background clutter and heavy occlusion efficiently. To handle this, discriminative information is collaborated to improve tracker stability and robustness. Collaborative trackers combining information from multiple models are not only computationally efficient, but have also shown superior performance during heavy occlusion scenarios.

REFERENCES

1. Kumar, B.S., M. Swamy, and M.O. Ahmad, Robust coding in a global subspace model and its collaboration with a local model for visual tracking. *Multimedia Tools and Applications*, 2020. **79**(7–8): pp. 4525–4551.
2. Yi, Y., Y. Cheng, and C. Xu, Visual tracking based on hierarchical framework and sparse representation. *Multimedia Tools and Applications*, 2018. **77**(13): pp. 16267–16289.
3. Jia, X., H. Lu, and M.-H. Yang, Visual tracking via coarse and fine structural local sparse appearance models. *IEEE Transactions on Image processing*, 2016. **25**(10): pp. 4555–4564.
4. Wen, H., X. Chen, and L. Zhou. A coarse-to-fine object tracking based on attention mechanism. in *Journal of Physics: Conference Series*. 2021. IOP Publishing.
5. Kong, J., Y. Ding, M. Jiang, and S. Li, Collaborative model tracking with robust occlusion handling. *IET Image Processing*, 2020. **14**(9): pp. 1701–1709.
6. Zhou, T., Y. Lu, and H. Di, Locality-constrained collaborative model for robust visual tracking. *IEEE Transactions on Circuits and Systems for Video Technology*, 2017. **27**(2): pp. 313–325.
7. Liu, C., P. Liu, W. Zhao, and X. Tang, Robust tracking and redetection: Collaboratively modeling the target and its context. *IEEE Transactions on Multimedia*, 2018. **20**(4): pp. 889–902.
8. Kumar, A., G.S. Walia, and K. Sharma, A novel approach for multi-cue feature fusion for robust object tracking. *Applied Intelligence*, 2020. **50**(10): pp. 3201–3218.
9. Kumar, A., G.S. Walia, and K. Sharma, Robust object tracking based on adaptive multicue feature fusion. *Journal of Electronic Imaging*, 2020. **29**(6): p. 063001.
10. Walia, G.S., H. Ahuja, A. Kumar, N. Bansal, and K. Sharma, Unified graph-based multicue feature fusion for robust visual tracking. *IEEE Transactions on Cybernetics*, 2019. **50**(6): pp. 2357–2368.
11. Walia, G.S., S. Raza, A. Gupta, R. Asthana, and K. Singh, A novel approach of multi-stage tracking for precise localization of target in video sequences. *Expert Systems with Applications*, 2017. **78**: pp. 208–224.
12. Xu, T., Z.-H. Feng, X.-J. Wu, and J. Kittler, Learning low-rank and sparse discriminative correlation filters for coarse-to-fine visual object tracking. *IEEE Transactions on Circuits and Systems for Video Technology*, 2019. **30**(10): pp. 3727–3739.
13. Kim, C., D. Song, C.-S. Kim, and S.-K. Park, Object tracking under large motion: Combining coarse-to-fine search with superpixels. *Information Sciences*, 2019. **480**: pp. 194–210.
14. Zhang, W., K. Song, X. Rong, and Y. Li, Coarse-to-fine UAV target tracking with deep reinforcement learning. *IEEE Transactions on Automation Science and Engineering*, 2018. **16**(4): pp. 1522–1530.
15. Li, D., G. Wen, and Y. Kuai, Collaborative convolution operators for real-time coarse-to-fine tracking. *IEEE Access*, 2018. **6**: pp. 14357–14366.
16. Zhong, B., B. Bai, J. Li, Y. Zhang, and Y. Fu, Hierarchical tracking by reinforcement learning-based searching and coarse-to-fine verifying. *IEEE Transactions on Image Processing*, 2019. **28**(5): pp. 2331–2341.

17. Li, D., G. Wen, Y. Kuai, and F. Porikli, Beyond feature integration: A coarse-to-fine framework for cascade correlation tracking. *Machine Vision and Applications*, 2019. **30**(3): pp. 519–528.
18. Wu, F., C.M. Vong, and Q. Liu, Tracking objects with partial occlusion by background alignment. *Neurocomputing*, 2020. **402**: pp. 1–13.
19. Zgaren, A., W. Bouachir, and R. Ksantini. Coarse-to-fine object tracking using deep features and correlation filters. in *Advances in Visual Computing: 15th International Symposium, ISVC 2020, San Diego, CA, USA, October 5–7, 2020, Proceedings, Part I 15*. 2020. Springer.
20. Han, G., J. Su, Y. Liu, Y. Zhao, and S. Kwong, Multi-stage visual tracking with Siamese anchor-free proposal network. *IEEE Transactions on Multimedia*, 2021.
21. Zeng, Y., B. Zeng, X. Yin, and G. Chen, SiamPCF: Siamese point regression with coarse-fine classification network for visual tracking. *Applied Intelligence*, 2022. **52**(5): pp. 4973–4986.
22. Horn, B.K. and B.G. Schunck, Determining optical flow. *Artificial Intelligence*, 1981. **17**(1–3): pp. 185–203.
23. Ma, M. and J. An, Combination of evidence with different weighting factors: A novel probabilistic-based dissimilarity measure approach. *Journal of Sensors*, 2015. **2015**.
24. Wu, Y., J. Lim, and M.-H. Yang. Online object tracking: A benchmark. in *Proceedings of the IEEE Conference on Computer Vision and Pattern Recognition*. 2013.
25. Kristan, M., A. Leonardis, J. Matas, M. Felsberg, R. Pflugfelder, L. Čehovin Zajc, ... A. Eldesokey. The sixth visual object tracking vot2018 challenge results. in *Proceedings of the European Conference on Computer Vision (ECCV) Workshops*. 2018.
26. Zhang, J., S. Ma, and S. Sclaroff. MEEM: Robust tracking via multiple experts using entropy minimization. in *Computer Vision–ECCV 2014: 13th European Conference, Zurich, Switzerland, September 6–12, 2014, Proceedings, Part VI 13*. 2014. Springer.
27. Bertinetto, L., J. Valmadre, S. Golodetz, O. Miksik, and P.H. Torr. Staple: Complementary learners for real-time tracking. in *Proceedings of the IEEE Conference on Computer Vision and Pattern Recognition*. 2016.
28. Mueller, M., N. Smith, and B. Ghanem. Context-aware correlation filter tracking. in *Proceedings of the IEEE Conference on Computer Vision and Pattern Recognition*. 2017.
29. Lazarevic-McManus, N., J. Renno, D. Makris, and G.A. Jones, An object-based comparative methodology for motion detection based on the F-measure. *Computer Vision and Image Understanding*, 2008. **111**(1): pp. 74–85.
30. Dou, J., Q. Qin, and Z. Tu, Robust visual tracking based on generative and discriminative model collaboration. *Multimedia Tools and Applications*, 2017. **76**(14): pp. 15839–15866.
31. Dai, M., S. Cheng, and X. He, Hybrid generative–discriminative hash tracking with spatio-temporal contextual cues. *Neural Computing and Applications*, 2018. **29**(2): pp. 389–399.
32. Duffner, S. and C. Garcia, Using discriminative motion context for online visual object tracking. *IEEE Transactions on Circuits and Systems for Video Technology*, 2016. **26**(12): pp. 2215–2225.

33. Bao, H., Y. Lu, H. Dai, and M. Lin, Collaborative tracking based on contextual information and local patches. *Machine Vision and Applications*, 2019. **30**: pp. 587–601.
34. Zhuang, B., L. Wang, and H. Lu, Visual tracking via shallow and deep collaborative model. *Neurocomputing*, 2016. **218**: pp. 61–71.
35. Tu, F., S.S. Ge, Y. Tang, and C.C. Hang, Robust visual tracking via collaborative motion and appearance model. *IEEE Transactions on Industrial Informatics*, 2017. **13**(5): pp. 2251–2259.
36. Liu, Y., F. Yang, C. Zhong, Y. Tao, B. Dai, and M. Yin, Visual tracking via salient feature extraction and sparse collaborative model. *AEU-International Journal of Electronics and Communications*, 2018. **87**: pp. 134–143.
37. Wang, Y., X. Luo, L. Ding, and S. Hu, Multi-task based object tracking via a collaborative model. *Journal of Visual Communication and Image Representation*, 2018. **55**: pp. 698–710.
38. Zhao, Z., L. Xiong, Z. Mei, B. Wu, Z. Cui, T. Wang, and Z. Zhao, Robust object tracking based on ridge regression and multi-scale local sparse coding. *Multimedia Tools and Applications*, 2020. **79**(1): pp. 785–804.
39. Kumar, A., R. Jain, V. A. Devi, & A. Nayyar, (Eds.). *Object Tracking Technology: Trends, Challenges, Impact, and Applications*, 2023. Springer.

Chapter 9

Deep learning-based visual tracking model

A paradigm shift

9.1 INTRODUCTION

DL-based tracking algorithms have attracted many researchers for robust target representation for efficient results. The DL networks CNN [1], ResNet [2], Siamese network [3], and LSTM [4] are exploited for tracking the trajectory of the target during tough tracking conditions. CNN can be employed for feature extraction and target classification from the background [5, 6]. Also, the features extracted from different layers of a DL network are integrated to generate efficient tracking results. It has been well accepted that the lower layers contain feature information rich in spatial data while the higher CNN layers are rich in discriminative semantic information [7, 8]. Hyper-features-based trackers integrate these features to obtain accurate tracking results. Trackers based on CNN feature extraction utilizing tracking by detection algorithms to address tracking problems. In this, the target is manually detected in the first frame. Based on the detected target, the deep learning model is trained to learn the features, and this is used to keep track of the target in future frames. Tracker learning and training can be done either online or offline depending on the complexity of the tracking architecture.

Recently, deep CNN-based networks, especially Siamese, are widely explored to provide advanced tracking solutions [3, 9–11]. Siamese networks are intensive in computations and have shown significant improvement in tracking outcomes. Due to this, the training and learning of Siamese networks for real-time tracking applications is not feasible. Many variants such as attentional transfer learning [12], fast attentional learning [3], residual learning [13], reinforcement learning [14], among others [11, 15] are integrated into the Siamese network to adopt a tradeoff between the training and the performance. Siamese networks consider tracking as a matching problem and compute the similarity between the target template and the reference template.

To address total darkness, complex occlusion, and target disappearance, the complementary vision information is integrated with multi-modal data such as thermal, IR, and depth in a deep learning network [16–19].

DOI: 10.1201/9781003456322-9

Multi-modal data is essential to ensure the reliability of the tracker when vision information fails to provide reliable tracking results. Complementary benefits of vision, IR, thermal, and depth sensors are explored to prevent tracking failures due to the limitations of vision sensors. Multi-modal information is widely used in deep learning trackers to propose trackers namely, DSiamMFT [19], Single-scale RGBD tracker [20], and many more [18, 21].

In this chapter, we have detailed the trackers based on deep learning networks. Deep learning networks can be used for robust feature extraction to provide stable trackers. Probabilistic tracking under the DL model using PF framework is explored to track a moving object. Deep networks-based Siamese trackers have shown significant improvement. Siamese-based trackers are complex in terms of computation, hence various Siamese network variants are explored to bridge the gap between the processing computations and the accuracy. Multi-model information in deep learning networks is integrated with RGB to expand the tracker's capabilities. The next section will discuss the various deep learning tracking frameworks.

9.2 DEEP LEARNING-BASED TRACKING FRAMEWORK

DL-based trackers are broadly categorized as probabilistic tracking and tracking by learning. These trackers utilize the DL network for extracting discriminating features suitable for stable tracking results. Table 9.1 shows the salient features along with the fusion and tracking strategy. The representative work is arranged in descending order of their published year. The details about the various tracking algorithms utilizing CNN for feature extraction are as follows.

9.2.1 Probabilistic deep convolutional tracking

To achieve better tracking results and robust feature extraction, trackers utilize deep learning networks. In this direction, the authors have utilized a pre-trained CNN network in the PF framework [1]. Handcrafted color features in HSV color space are integrated with deep features to obtain a robust appearance model by reducing the tracker's computations. Automatic target localization is achieved by Bayesian estimation. In ref. [22], the authors have extracted deep features from CNN and used a hybrid gravitational search algorithm to prevent particle loss in the PF framework. Particle swarm optimization explores past information for fast convergence of tracking algorithms.

Further, Cao et al. [23] have integrated intensity and CNN features to extract the spatio-temporal context information. The confidence index from the dynamic confidence feature map is used for fine-tuning and online model updates. The authors have selected convolutional layers adaptively to obtain robust features [6]. To prevent the tracker's update from incorrect tracking

Table 9.1 Description of representative work under DL-tracking framework

SN	*References*	*Year*	*DL architecture*	*Tracking strategy*	*Fusion strategy*	*Summary*
1.	Kang et al. [22]	2018	CNN-based feature extraction	Swarm intelligence-based probabilistic method	—	Integrated deep features to overcome the limitation of handcrafted HSV features.
2.	Qian et al. [1]	2018	CNN for deep feature extraction	PF-based probabilistic method	Weighted fusion	Integrate deep features, HSV color, and motion information in the PF framework to reduce the number of particles for accurate estimation.
3.	Cao et al. [23]	2018	CNN for feature representation	Dynamic mapping of the confidence map	Summation of the confidence map	Dynamic training of confidence map for accurate tracker's location during occlusion.
4.	Gan et al. [24]	2018	Motion-guided CNN feature	Tracking by detection	—	Integrate spatial color and temporal optical flow in an end-to-end CNN-trained network.
5.	Tang et al. [6]	2019	CNN-based feature extraction	Feature map extraction from different layers of CNN and reselection	—	Adaptive selection of features from different layers of CNN along with verification mechanism to prevent false tracking results.
6.	Danelljan et al. [25]	2019	CNN for feature learning	Tracking by detection	Fourier interpolation method	Fusion of deep features with handcrafted features to obtain efficient tracking results.
7.	Lu et al. [8]	2019	CNN for feature extraction	Multi-task learning using end-to-end network	—	Consisting of deep feature extraction network and deep regression network for the tracker's online update.
8.	Li et al. [2]	2019	Spatial and temporal CNN-based feature extraction	Tracking by detection	Kernelized CF-based fusion	Integrate different-size feature maps using the fusion method for accurate localization of the target.

(*Continued*)

Table 9.1 (Continued)

SN	*References*	*Year*	*DL architecture*	*Tracking strategy*	*Fusion strategy*	*Summary*
9.	Wang et al. [26]	2019	CNN-based feature extraction	Multi-layer CNN-based tracking	Weighted fusion	Selective and online update of the translation filter for effective tracking results during abrupt target deformations.
10.	Zhang et al. [27]	2020	CNN for deep feature extraction	CF-based feature map construction	—	Multiscale estimation to estimate the target scaling features to handle scale variation during tracking.
11.	Nousi et al. [5]	2020	CNN for feature extraction	Histogram from deep features from different layers	Similarity-based concatenation	End-to-end learnable architecture with offline training to improve computational efficiency.
12.	Yuan et al. [28]	2020	CNN	Adaptive structural CNN	Adaptive weighting strategy	Target structural pattern integrated with original filter layer to obtain accurate tracking.
13.	Zhao et al. [29]	2021	Fully CNN-based mutual learning	TLD	KL divergence	Fully CNN-based feature extraction for end-to-end training for improving the tracker's performance.
14.	Gurkan et al. [30]	2021	Mask R-CNN	TLD	—	Long-term tracking along with adaptive detection without additional training.
15.	Fang and Liu [31]	2021	CNN	TLD	Multiscale feature fusion	Discriminative representation to perform scale invariant tracking of small objects.
16.	Lu et al. [10]	2023	CNN for feature extraction	Fully CNN with the attentional mechanism	Feature pyramid network	Utilize heterogeneous kernels to reduce the tracker's computations by improving the speed.

Note: The work is arranged in the ascending order of their published year.

prediction, a verification mechanism is adopted. The maximum filter response score is used for updating the target's scale. To generate powerful features, feature histograms are extracted from the features from various layers of CNN [5]. Feature histogram from different layers is integrated to avoid the covariance shift between the features using Bag-of-Features pooling. However, the authors have proposed a multi-layer CNN method for adaptive scale estimation [26]. The target's accurate localization is performed by a weighted fusion of response from scale and estimation filters. The filters are selectively updated to keep the tracker adaptive to the tracking variations and prevent drift. Nai et al. [32] have extracted high-dimensional CNN features to extract the target from the background. Features selection strategy is incorporated to improve the discriminative and generalizability of the tracker during tracking challenges. In ref. [28], the authors have utilized the structural convolutional local filter layer to capture the structural pattern of the target. Peak sidelobe ratio and Laplacian distribution are applied with an adaptive weighting strategy for stable tracking results. In summary, probabilistic tracking methods are robust in performance during complex environmental variations. These trackers generate discriminative features to strengthen the target's appearance model for accurate localization.

9.2.2 Tracking by detection deep convolutional tracker

In this direction, the authors have utilized residual learning in a deep ResNet framework [2]. The tracker is trained offline to obtain multiscale features. The spatial and temporal stream is fused and fine-tuned in the multiscale ResNet. The outputs are fed to the various kernelized CFs to obtain the scaling variation adaptively. The correlation between the consecutive video frames is modeled using a motion-guided CNN [24]. The spatial and temporal features, RGB and optical flow, are processed in end-to-end trainable networks. Optical flow captures the pixel motion in the adjacent video frames to obtain accurate target localization. The model is trained offline to reduce the computational load. Zhao et al. [29] have introduced mutual learning in a fully connected CNN. The typical TLD tracker and CF tracker are integrated to obtain robust tracking results. The tracker is trained offline to reduce computations, but training the objective, optimization, and loss function in a mutual learning environment is tedious. However, authors have fused image patch information from various contexts using mutual learning [31]. Multiscale feature pyramid fusion is used to address the scale variations in small-size targets. The weighted cross-entropy loss function is used to handle object classification and tracking. In ref. [30], tracking and detection are performed using FPN-ResNet101 in Mask R-CNN jointly. Similarity-based temporal matching is utilized for modeling the tracker's appearance model. LBP histogram is incorporated to monitor the target presence in the scene in long-term tracking. Target size and scale are handled by incorporating scale adaptive region proposal network. In summary, tracking by detection trackers is more discriminative, and separates targets from the background efficiently. Mutual learning is a concept utilized by these trackers to fuse the results from the various tracking strategies. The tracker is trained online and offline to maintain the computational complexity of the tracker.

9.3 HYPER-FEATURE-BASED DEEP LEARNING NETWORKS

Hyper-features are those extracted from various layers of a deep neural network and integrate these features to obtain powerful discriminative features. These features are considered to be robust and efficient for target representation in complex scenarios. Siamese trackers are some of the most popular trackers under this category. However, these trackers utilize deep neural networks for feature representation and have quite complex architecture. They integrate other technologies such as residual learning [13], cross-residual learning [33], attentional transfer learning [12], among others [11, 34], to reduce computations and enhance efficiency. Also, hyper-features can be extracted from various other deep neural networks. This category of trackers utilizes RNN (Recurrent Neural Network) [35], LSTM (Long Short-Term Memory network) [4], and other networks [36] for feature extraction. Table 9.2 shows the various Siamese trackers, and trackers with specialized networks with their salient features. The details of these Siamese trackers are as follows.

9.3.1 Siamese network-based trackers

Siamese trackers are not efficient enough to differentiate the target from the background in the presence of distractors even if both are of different color. To address this, the authors have combined the handcrafted color feature with the high-level semantic features extracted from the fully convolutional network [7]. Color feature maps and CNN feature maps are integrated to generate a unified feature map. Both feature maps are adjusted to the same size based on multilevel thresholds to eliminate the noise in the feature map. However, to reduce the distractors in long-term tracking, two-stage tracking is proposed by Xuan et al. [11]. A Siamese-based verification network is utilized for better identification of the tracking in presence of environmental distractors. The candidate generation network classifies the anchor maps either as target or background. The distractor reduction method refines the networks by reducing the impact of negative anchors in the training data set. In ref. [12], the target and background are modeled by using an attentional online-based update strategy. Decision-making-based on similarity is

Table 9.2 Description of representative work under hyper-features-based DL tracking framework

SN	*References*	*Year*	*DL architecture*	*Tracking strategy*	*Fusion strategy*	*Summary*
1.	Wang et al. [35]	2017	CNN and RNN-based feature extraction	LSTM-based prediction for tracking	Dynamic weighted motion model	Dynamic weighted model combines with random sampler to predict difficult object trajectory.
2.	Gao et al. [7]	2020	Fully convolutional Siamese network	Color score map and CNN score map integration	Score level fusion	Effective area cropping along with Gaussian smoothing to effectively obtain the target from the fused score maps.
3.	Liu et al. [4]	2020	LSTM network	Probabilistic PF	—	Object detection followed by tracking with end-to-end fully connected LSTM-PF.
4.	Zhang et al. [13]	2021	Siamese network(normal features)	Spatial attention-guided residual learning	Weighted adaptive fusion based on residual learning	Loss function to balance the mismatch between the easy and hard samples for effective training of the network.
5.	Xuan et al. [11]	2021	Siamese network	Two-stage tracking	—	Candidate generation network and verification network to enhance network capability.
6.	Noor et al. [34]	2021	Fully convolutional SiamMask	Adaptive feature mask	—	Automatic object detection along with end-to-end network for adaptive object tracking.
7.	Wu et al. [33]	2021	Deep Siamese network	Cross-residual learning	Weighted summation	Combined loss function to entangle the outputs from two branches of the Siamese network for learning effective features.
8.	Zhao et al. [14]	2021	Siamese regression network	Reinforcement learning	—	Online learning with anchor network to adapt tracker with changing target variations.

(*Continued*)

Table 9.2 (Continued)

SN	*References*	*Year*	*DL architecture*	*Tracking strategy*	*Fusion strategy*	*Summary*
9.	Huang et al. [12]	2022	Siamese network	Attentional transfer learning	Mutual compensation mechanism	Online update of the tracker using attentional learning for tracker's stability during changing backgrounds.
10.	He and Sun [9]	2022	Siamese network	Context-aware tracking	Weighted fusion	Global learning module to generate relationship between template and search.
11.	Yang et al. [15]	2022	Siamese network	Anchor free tracking	Weighted multi-layer feature fusion	Added a modified corner layer to extract corners from the BB for discriminating targets from the background.
12.	Liu et al. [16]	2022	Dual deep Siamese network	Target-aware module	Linear adaptive fusion	Online template update to improve processing and generate hyperspectral objects to reduce the data required for training the network.
13.	Choi et al. [37]	2022	Siamese network	Meta-learning	—	Dynamic generation of hyperparameters to enhance the tracker's speed with adaptive learning of the parameters.
14.	Qin et al. [3]	2022	Siamese network	Fast attention network	—	Attentional module to determine accurate target position for robust tracking results.
15.	Ge et al. [36]	2023	Dual Attention-aware Network	CNN and Attention Cooperative	Depth Cross-Correlation network	Integrate global and local features at different resolutions to improve the discriminative ability of the tracker.

Note: The works are arranged in the ascending order of their year of publication.

utilized by using attentional transfer learning in the Siamese network. The discriminative ability of the tracker is enhanced by using instance transfer learning-based CF during online tracking. Zhang et al. [13] have proposed spatial attention-guided residual learning in Siamese networks to extract the target's contextual information for efficient tracking. Handcrafted features and attention-weighted features are integrated to prevent the tracker's degradation as well as overfitting. The imbalance between the hard and easy samples is addressed using a joint loss function.

To improve the tracking performance by addressing the computation limitations of Siamese trackers, fast attention networks are embedded in the Siamese networks [3]. Attention networks reduce computations by eliminating redundant operations. Also, multi-layer perceptron reduces the hyperparameters in Siamese architecture. The similarity between the target and the search area is calculated using the attention model to obtain an accurate target position. However, the similarity between the target and the search area is determined by considering the contextual association between them [9]. Cascade cross-correlation module refines the features information to improve accuracy. Zhao et al. [14] have proposed reinforcement learning in Siamese trackers to effectively capture the target appearance. Online and offline training with location module subnetwork is executed for accurate target localization. To improve the accuracy and efficiency of the Siamese networks, region proposal networks are used by [15]. Region proposal network improves the tracker's design by selecting the number, scale, and aspect ratios of the aspect boxes in end-to-end offline trained networks. The corner pooling layer is to estimate the target using a pair of corner predictions.

On the other hand, a dual deep Siamese network is designed for tracking hyperspectral objects using a target-aware module [16]. Pre-trained RGB network is used to extract the spectral information for target state estimation. The tracker is trained with fewer videos and achieved high accuracy in hyperspectral videos. Similarly, cross-residual learning is proposed in the Siamese networks to reduce the number of training videos [33]. The combined loss function is used for matching and classification tasks. Instance-specific information learns the target features at different nodes for compact representation. Noor et al. [34] have proposed unsupervised SiamMask to perform the detection and tracking jointly. Detectron2 is for the automatic detection of the target and the fully convolutional Siamese network is for tracking the object. Authors have proposed adaptive meta-learning for generating hyperparameters to achieve the tracker's fast initialization and online update [37]. Meta-learning keeps a record of the previously gathered knowledge and utilizes it for the adaptive update of the tracker during tracking variations. In summary, Siamese trackers are highly accurate and robust for tracking objects in a tedious environment. However, the complexity of their networks has encouraged researchers to optimize the performance for real-time speed.

9.3.2 Specialized deep network-based trackers

Apart from Siamese networks, other deep neural networks such as RNN and LSTM have gained attraction to achieve stable tracking performance. In this direction, the authors have proposed LSTM-PF to track the target in complex videos [4]. A set of weighted particles approximates the target location and updates the state in LSTM gating structure as per Bayesian rules. The tracker is trained offline and online for fine-tuning with fewer training samples. However, RNN is used for trajectory prediction for estimating the target motion [35]. In addition, the LSTM model extracts features and center locations of the tracked target with sequential visual information from the annotated trajectories. To handle the tedious trajectory prediction, a dynamic weighted motion model along with a random sampler is used. On the other hand, Ge et al. [36] have proposed dual attention-aware mechanism in a parallel network for accurate target boundary detection. CNN and attention cooperative processing modules are used for robust feature extraction. Dual attention-aware network is enhanced by a feature pyramid structure for feature enhancement. In summary, deep learning networks are suitable for integrating local and global information for high-tracking precision results. The tedious target trajectory prediction is performed in complex videos using a highly deep and complex neural network.

9.4 MULTI-MODAL BASED DEEP LEARNING TRACKERS

Tracking an object becomes tedious when the visible information captured from the sensor is reliable. During poor illumination, darkness, and complex occlusion, RGB data is inefficient to provide robust tracking results. To prevent tracker drift under these typical conditions, multi-modal information is integrated with a deep learning network. Precisely, RGB data is integrated with the information captured from the specialized sensors extracting the thermal profile and the depth information from the target. The salient features of the representative work exploiting multi-modal information in a deep neural network are shown in Table 9.3.

To address the limitations of the RGB trackers, complementary information from thermal IR images is fused with RGB images in a dynamic multilayer Siamese network [19]. Response maps generated from the different layers of an RGB network and the IR network are fused using elementwise fusion to localize the target. However, the authors have proposed a quality-aware fusion module for aggregating the bilinear pooling features extracted from a cross-layer bilinear pooling network [17]. Semantic deep features, texture features, and thermal modalities are calibrated using the channel attention mechanism before the feature fusion for boosting performance. Also, quality-aware integration of RGB data and thermal modality using an attention-based mechanism is proposed by Tu et al. [38]. Multi-margin

Table 9.3 Description of representative work under multi-modal DL-tracking framework

SNO	*Reference*	*Year*	*DL architecture*	*Tracking strategy*	*Fusion strategy*	*Summary*
1.	Zhang et al. [19]	2020	Siamese network	Dynamic multi-layer fusion of deep features	Elementwise fusion strategy	Feature response maps from different layers of deep neural network fuse in a multi-feature fusion strategy.
2.	Xiao et al. [20]	2021	Siamese network	3D local view field	Weighted average	Adaptive update strategy to handle full occlusion and partial occlusion by combining RGB and depth information.
3.	Lu et al. [40]	2021	MANet++	Multi-Adapter Network	Quality-aware dynamic fusion	Multiple target representations to extract modality-specific features and reduce computational complexity.
4.	Liu et al. [41]	2021	Siamese network	Multilevel similarity network	Relative entropy-based ensemble subnetwork	Complementary information to enhance the tracker's discriminative capability to cater to distractors.
5.	Xu et al. [17]	2022	CBPNET	Cross-Layer Bilinear Pooling	Quality-aware feature fusion	Channel attentional mechanism to generate discriminative features and quality-aware fusion to suppress the background noise.
6.	Guo et al. [39]	2022	Dual Siamese network	Joint modal channel attention	Response-level fusion	Weight distribution with position maps fusion using proposal subnetworks for improving network performance.
7.	Tang et al. [21]	2022	DFAT	Modulation of RPN block with adaptive weights	Adaptive decision-level fusion	The dynamic weighting of RGB and TIR along with adaptive linear template update.
8.	Tu et al. [38]	2022	RT-MDNet	Multi-Margin Metric Learning	Attention-based fusion	Preserve structural loss from hard samples of both RGB and thermal modalities for improving sample classification.
9.	Zhang et al. [18]	2023	Self-supervised Siamese network	Cross-Input Consistency	Feature-level fusion	Adopt a re-weighting scheme to recalculate the loss for improved training of the tracker.

Note: The work is arranged in the ascending order of their published year.

structural loss of hard samples of the target is preserved in the training stage. Hard samples are optimized to obtain the accurate target position. In ref. [39], the authors propose response-level fusion for integrating the RGB features and thermal modalities. Dual Siamese subnetworks are fused using the regional proposal subnetworks for predicting the target's position maps.

Weight distribution is modeled in the feature extraction stage for designing the joint modal attention module. Zhang et al. [18] have utilized a reweighting strategy to train the tracker from low-quality samples. Cross-input consistency loss between the two pairs of inputs is constructed using a Siamese correlation filter network for effective tracking results.

Apart from thermal modality, depth maps are also integrated with RGB information to construct a robust tracker [20]. The BB is automatically adjusted to achieve stable tracking results using a single-scale Siamese network. Target depth along with 3D visual filter maintains the tracker's performance by eliminating noise during background clutter and occlusion. Backtracking is performed to recover the target from heavy occlusion by updating the best template. Target local structural and global semantic similarities are modeled by learning the target's weight adaptively [41]. These similarities are integrated using an entropy-based ensemble subnetwork. The integrated features generate features that are robust during various tracking challenges. For accurate tracking, multiple modalities are integrated using different fusion techniques namely, pixel-level fusion, feature-level fusion, and decision-level fusion explored by [21]. Decision-level fusion can generate discriminative features in comparison with other fusion techniques. Target templates are adaptively updated to address tracking challenges during long-term tracking [42]. In summary, multi-modality-based trackers are robust to distractors during tracking which RGB-based trackers are not able to address efficiently [43].

9.5 SUMMARY

In this chapter, we have discussed the salient features of various multi-modal based trackers utilizing end-to-end deep neural architecture, and proposed that all-day and all-weather tracking solutions, multi-modal sensors data (thermal, IR), and depth are integrated with RGB complementary features. Modality-specific features and quality-aware fusion modules are utilized to generate the discriminative features. Various types of fusion techniques are explored in multi-modal trackers to facilitate the adaptive update of the target template. Deep learning networks in the RGBT have obtained high-precision results with accurate localization of the target. Some networks utilize a single network for extracting the various modalities. However, recent works utilize the dual network for extracting the different modalities so that effective features can be captured efficiently.

The main challenge of exploring the multi-modality in the deep neural network is to maintain the computational complexity for real-time speed. For this, attention mechanisms, context-aware features, and quality-aware fusion techniques are widely explored. Also, adaptive template update schemes are utilized to reduce noise in the tracker during complex background clutter and total occlusion.

REFERENCES

1. Qian, X., L. Han, Y. Wang, and M. Ding, Deep learning assisted robust visual tracking with adaptive particle filtering. *Signal Processing: Image Communication*, 2018. **60**: pp. 183–192.
2. Liu, B., Q. Liu, T. Zhang, and Y. Yang, MSSTResNet-TLD: A robust tracking method based on tracking-learning-detection framework by using multiscale spatio-temporal residual network feature model. *Neurocomputing*, 2019. **362**: pp. 175–194.
3. Qin, L., Y. Yang, D. Huang, N. Zhu, H. Yang, and Z. Xu, Visual tracking with Siamese network based on fast attention network. *IEEE Access*, 2022. **10**: pp. 35632–35642.
4. Liu, Y., J. Cheng, H. Zhang, H. Zou, and N. Xiong, Long short-term memory networks based on particle filter for object tracking. *IEEE Access*, 2020. **8**: pp. 216245–216258.
5. Nousi, P., A. Tefas, and I. Pitas, Dense convolutional feature histograms for robust visual object tracking. *Image and Vision Computing*, 2020. **99**: p. 103933.
6. Tang, F., X. Lu, X. Zhang, L. Luo, S. Hu, and H. Zhang, Adaptive convolutional layer selection based on historical retrospect for visual tracking. *IET Computer Vision*, 2019. **13**(3): pp. 345–353.
7. Gao, T., N. Wang, J. Cai, W. Lin, X. Yu, J. Qiu, and H. Gao, Explicitly exploiting hierarchical features in visual object tracking. *Neurocomputing*, 2020. **397**: pp. 203–211.
8. Lu, X., F. Tang, H. Huo, and T. Fang, Learning channel-aware deep regression for object tracking. *Pattern Recognition Letters*, 2019. **127**: pp. 103–109.
9. He, X. and Y. Sun, SiamBC: Context-related Siamese network for visual object tracking. *IEEE Access*, 2022. **10**: pp. 76998–77010.
10. Lu, Z., Y. Bian, T. Yang, Q. Ge, and Y. Wang, A new Siamese heterogeneous convolutional neural networks based on attention mechanism and feature pyramid. *IEEE Transactions on Cybernetics*, 2023.
11. Xuan, S., S. Li, Z. Zhao, L. Kou, Z. Zhou, and G.-S. Xia, Siamese networks with distractor-reduction method for long-term visual object tracking. *Pattern Recognition*, 2021. **112**: p. 107698.
12. Huang, B., T. Xu, Z. Shen, S. Jiang, B. Zhao, and Z. Bian, Siamatl: Online update of Siamese tracking network via attentional transfer learning. *IEEE Transactions on Cybernetics*, 2022. **52**(8): pp. 7527–7540.
13. Zhang, D., Z. Zheng, M. Li, and R. Liu, CSART: Channel and spatial attention-guided residual learning for real-time object tracking. *Neurocomputing*, 2021. **436**: pp. 260–272.

14. Zhao, F., T. Zhang, Y. Song, M. Tang, X. Wang, and J. Wang, Siamese regression tracking with reinforced template updating. *IEEE Transactions on Image Processing*, 2021. **30**: pp. 628–640.
15. Yang, K., Z. He, W. Pei, Z. Zhou, X. Li, D. Yuan, and H. Zhang, SiamCorners: Siamese corner networks for visual tracking. *IEEE Transactions on Multimedia*, 2022. **24**: pp. 1956–1967.
16. Liu, Z., X. Wang, Y. Zhong, M. Shu, and C. Sun, SiamHYPER: Learning a hyperspectral object tracker from an RGB-based tracker. *IEEE Transactions on Image Processing*, 2022. **31**: pp. 7116–7129.
17. Xu, Q., Y. Mei, J. Liu, and C. Li, Multi-modal cross-layer bilinear pooling for RGBT tracking. *IEEE Transactions on Multimedia*, 2022. **24**: pp. 567–580.
18. Zhang, X. and Y. Demiris, Self-supervised RGB-T tracking with cross-input consistency. arXiv preprint arXiv:2301.11274, 2023.
19. Zhang, X., P. Ye, S. Peng, J. Liu, and G. Xiao, DSiamMFT: An RGB-T fusion tracking method via dynamic Siamese networks using multi-layer feature fusion. *Signal Processing: Image Communication*, 2020. **84**: p. 115756.
20. Xiao, F., Q. Wu, and H. Huang, Single-scale Siamese network-based RGB-D object tracking with adaptive bounding boxes. *Neurocomputing*, 2021. **451**: pp. 192–204.
21. Tang, Z., T. Xu, H. Li, X.-J. Wu, X. Zhu, and J. Kittler, Exploring fusion strategies for accurate RGBT visual object tracking. arXiv preprint arXiv:2201.08673, 2022.
22. Kang, K., C. Bae, H.W.F. Yeung, and Y.Y. Chung, A hybrid gravitational search algorithm with swarm intelligence and deep convolutional feature for object tracking optimization. *Applied Soft Computing*, 2018. **66**: pp. 319–329.
23. Cao, Y., H. Ji, W. Zhang, and F. Xue, Learning spatio-temporal context via hierarchical features for visual tracking. *Signal Processing: Image Communication*, 2018. **66**: pp. 50–65.
24. Gan, W., M.-S. Lee, C.-H. Wu, and C.-C.J. Kuo, Online object tracking via motion-guided convolutional neural network (MGNet). *Journal of Visual Communication and Image Representation*, 2018. **53**: pp. 180–191.
25. Danelljan, M., G. Bhat, S. Gladh, F.S. Khan, and M. Felsberg, Deep motion and appearance cues for visual tracking. *Pattern Recognition Letters*, 2019. **124**: pp. 74–81.
26. Wang, X., Z. Hou, W. Yu, Z. Jin, Y. Zha, and X. Qin, Online scale adaptive visual tracking based on multilayer convolutional features. *IEEE Transactions on Cybernetics*, 2019. **49**(1): pp. 146–158.
27. Zhang, J., X. Jin, J. Sun, J. Wang, and A.K. Sangaiah, Spatial and semantic convolutional features for robust visual object tracking. *Multimedia Tools and Applications*, 2020. **79**(21): pp. 15095–15115.
28. Yuan, D., X. Li, Z. He, Q. Liu, and S. Lu, Visual object tracking with adaptive structural convolutional network. *Knowledge-Based Systems*, 2020. **194**: p. 105554.
29. Zhao, H., G. Yang, D. Wang, and H. Lu, Deep mutual learning for visual object tracking. *Pattern Recognition*, 2021. **112**: p. 107796.
30. Gurkan, F., L. Cerkezi, O. Cirakman, and B. Gunsel, TDIOT: Target-driven inference for deep video object tracking. *IEEE Transactions on Image Processing*, 2021. **30**: pp. 7938–7951.

31. Fang, J. and G. Liu, Visual object tracking based on mutual learning between cohort multiscale feature-fusion networks with weighted loss. *IEEE Transactions on Circuits and Systems for Video Technology*, 2021. **31**(3): pp. 1055–1065.
32. Nai, K., Z. Li, and H. Wang, Learning channel-aware correlation filters for robust object tracking. *IEEE Transactions on Circuits and Systems for Video Technology*, 2022.
33. Wu, F., T. Xu, J. Guo, B. Huang, C. Xu, J. Wang, and X. Li, Deep Siamese cross-residual learning for robust visual tracking. *IEEE Internet of Things Journal*, 2021. **8**(20): pp. 15216–15227.
34. Noor, S., M. Waqas, M.I. Saleem, and H.N. Minhas, Automatic object tracking and segmentation using unsupervised SiamMask. *IEEE Access*, 2021. **9**: pp. 106550–106559.
35. Wang, L., L. Zhang, and Z. Yi, Trajectory predictor by using recurrent neural networks in visual tracking. *IEEE Transactions on Cybernetics*, 2017. **47**(10): pp. 3172–3183.
36. Ge, H., S. Wang, C. Huang, and Y. An, A visual tracking algorithm combining parallel network and dual attention-aware mechanism. *IEEE Access*, 2023.
37. Choi, J., S. Baik, M. Choi, J. Kwon, and K.M. Lee, Visual tracking by adaptive continual meta-learning. *IEEE Access*, 2022. **10**: pp. 9022–9035.
38. Tu, Z., C. Lin, W. Zhao, C. Li, and J. Tang, M 5 l: Multi-modal multi-margin metric learning for RGBT tracking. *IEEE Transactions on Image Processing*, 2022. **31**: pp. 85–98.
39. Guo, C., D. Yang, C. Li, and P. Song, Dual Siamese network for RGBT tracking via fusing predicted position maps. *The Visual Computer*, 2022. **38**(7): pp. 2555–2567.
40. Lu, A., C. Li, Y. Yan, J. Tang, and B. Luo, RGBT tracking via multi-adapter network with hierarchical divergence loss. *IEEE Transactions on Image Processing*, 2021. **30**: pp. 5613–5625.
41. Liu, Q., X. Li, Z. He, N. Fan, D. Yuan, and H. Wang, Learning deep multi-level similarity for thermal infrared object tracking. *IEEE Transactions on Multimedia*, 2021. **23**: pp. 2114–2126.
42. Kumar, A., G.S. Walia, and K. Sharma (2020). Recent trends in multicue based visual tracking: A review. *Expert Systems with Applications*, **162**: p. 113711.
43. Kumar, A., R. Jain, V. A. Devi, & A. Nayyar, (Eds.). *Object Tracking Technology: Trends, Challenges, Impact, and Applications*, 2023. Springer.

Chapter 10

Correlation filter-based visual tracking model

Emergence and upgradation

10.1 INTRODUCTION

In recent years, correlation filter (CF)-based trackers have gained popularity for tracking objects. CF-based trackers are more robust and efficient in comparison to traditional trackers. The tracker's dynamic model is stable and effective in the presence of environmental variations by adaptive feature learning from the target. Generally, CF tracker algorithms perform computations in the frequency domain to reduce processing costs [1]. However, a robust appearance model invariant to tracking challenges namely scale variations, fast motion, occlusion, and background clutters is still a challenge. Figure 10.1 illustrates the various categories of CF-based tracking frameworks.

CF-based tracking algorithms are categorized into various classes. Conventional CF-based trackers are categorized either as context-aware CF trackers [2, 3] or part-based CF trackers [4, 5]. Context includes the target as well as the surrounding background. CF-based trackers consider them as a whole and perform tracking accordingly. In context-aware tracking, CF-based trackers utilize both information to enhance the tracker's discriminability power. However, part-based CF trackers perform tracking in parts. Initially, local parts of the target are tracked by local filters and then combined to obtain the global tracking results. Under this, kernelized CF-based trackers are very popular [6, 7]. But these trackers tend to drift during tedious tracking situations. To address this, spatial regularization-based CF trackers are proposed [8, 9]. These trackers extract spatial and semantic features to develop a robust appearance model sparsity is introduced in the dynamic model to reduce noise and enhance the tracker's robustness in presence of tracking challenges.

To improve the tracking performance, CF-based trackers are explored with deep neural networks such as CNN and Siamese networks [10–13]. Trackers under this category extract high-level semantic features and low-level spatial features to develop a robust tracking model. Tracker complexity can be addressed by attentional mechanism and offline training of the

DOI: 10.1201/9781003456322-10

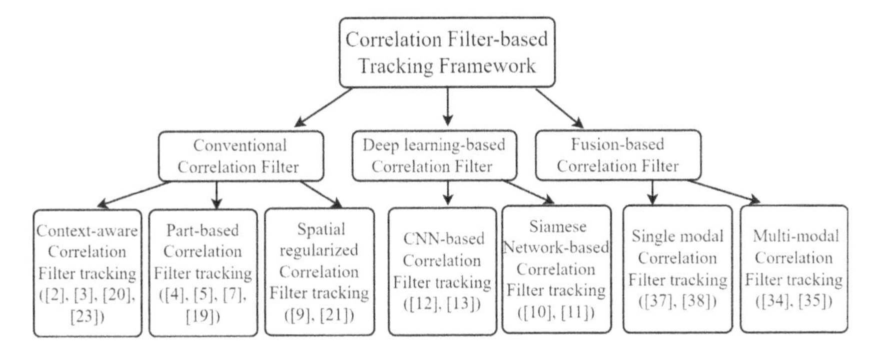

Figure 10.1 Illustration of various categories of correlation filer-based tracking.

trackers for robust feature learning. Dimensionality reduction methodology is also proposed to reduce the tracker's computations.

To enhance computational power and the tracker's robustness, fusion-based CF trackers are proposed [14–16]. Under this, trackers combine handcrafted features, deep features, and multi-modal features. Handcrafted features include color naming, HOG, and LBP extracted from vision sensors. Deep features are extracted from different layers of CNN networks that include either VGG-16 or VGGM. Multi-modal extracts target features from the specialized sensor such as depth. Generally, handcrafted features-based CF trackers offer real-time tracking solutions but are unstable due to less discriminative features. On the other hand, deep features are more powerful and efficient to develop an efficient tracker, but they are computationally complex due to a large number of parameters. Also, the vision sensor-based CF trackers are not able to handle severe occlusion and complex tracking challenges. Hence, handcrafted features, deep features, and multi-modal features are fused not only to compensate for each other limitations but also to develop robust tracking solutions. The next section will discuss the various CF-tracking algorithms.

10.2 CORRELATION FILTER-BASED TRACKING FRAMEWORK

Conventional CF-based tracking frameworks are categorized as context-aware CF trackers, part-based CF trackers, and spatially regularized based CF trackers. Table 10.1 shows the salient features of the representative work under each category. CF-based tracking frameworks develop robust appearance models by extracting the appropriate features from the target. The feature extraction process should extract discriminative features by maintaining the tracker's processing power. The details about the context-aware CF trackers are as follows.

Table 10.1 Description of representative work under CF-based tracking framework

SN	*References*	*Year*	*Algorithm*	*Feature*	*Update strategy*	*Summary*
1.	Bai et al. [17]	2018	Template matching	HOG and color naming	Data-driven based update	Adaptive scale scheme to deal with the fixed template size of CF filter.
2.	Li et al. [6]	2018	KCF	HOG and region covariance	Correlation response-based adaptive update	Perform re-detection by Gaussian constraint in correlation response with scale adaptation method.
3.	Li and Yang [5]	2019	Part-based tracking	Holistic and reliable local feature	Adaptive feedback-based update	Reliable parts are weighted to vote for the rough localization of the target.
4.	Li et al. [18]	2019	Target-aware tracking	HOG and color histogram	Adaptive linear update	Iterative optimization for efficient training of CF.
5.	Liu et al. [2]	2019	Context-based tracking	HOG and color names	Threshold-based adaptive update	Context pyramid in the context adaptive strategy for preventing background learning of tracker.
6.	Xu et al. [3]	2019	Context-based tracking	HOG and Color names	Interpolation-based update	Combine target information with background context to improve the tracker's discriminability.
7.	Wang et al. [19]	2019	Hybrid part-based tracking	Local kernel	Online threshold-based update	Peak-to-sidelobe ratio to measure the tracking result's reliability.
8.	Lian [7]	2020	Part-based tracking	Color names	Weighted update	Perform tracking by dividing the target into four parts by computing scale factors.
9.	Elayaperumal and Joo [20]	2021	Context-based tracking	Color names, HOG, gray, and CNN features	Weighted update	Extracts context information from spatial variations for the target's accurate localization.
10.	Zhang et al. [21]	2021	Sparse response regularization	–	–	Sparse regularization to improve tracking results during long-term tracking.

11.	Barcellos et al. [4]	2021	Part-based tracking	Deep features	Linear interpolation-based update	Hyper-features extract high-level and lower-level features efficiently to capture the target's fine details.
12.	Feng et al. [8]	2021	Spatial-temporal regularization	Gray, HOG, and color names	Measurement value-based update	Stride length control to limit the maximum amplitude of the output state.
13.	Zhu et al. [22]	2022	Attribute aware tracking	Spatial sparsity	Adaptive online update	Reduce the irrelevant and inconsistent channels by representing discriminative attributes.
14.	Feng et al. [23]	2022	Context-aware tracking	HOG, Color name, histogram of local intensity	Threshold-based update	Group feature fusion by the weighted sum of normalized multi-feature response maps.
15.	Chen et al. [24]	2022	Gaussian-like function label	HOG, color names, and deep features from ResNet-34	Incremental update	Aspect ratio distribution-based tracking to facilitate the robust training of the tracker.
16.	Cao et al. [25]	2022	Gaussian scale mixture model	31-channel HOG, 10-channel color names, and deep features from VGG-M	Lagrangian multiplier-based iterative update	Structured GSM to extract spatial correlations among CFs to improve tracking results.
17.	Zhang et al. [9]	2023	Sparse spatially regularization	HOG and deep features from VGG16 and VGGM2048	Online linear update	Elastic net regression-based sparsity and ADMM-based optimization.

The works are arranged in the ascending order of year of publication.

10.2.1 Context-aware correlation filter-based trackers

Context-aware CF trackers focus to capture the target appearance as well as the nearby background to propose robust tracking solutions. In this direction, the authors have proposed multi-level context adaptive tracking [2]. Target's features are jointly learned from the different contexts in the CF framework. Potential background information is also considered by the weighted 3D spatial window during the tracker's learning stage. Context adaptive strategy is to reduce the impact of background information as well as to track the different sizes of targets and samples. Xu et al. [3] have jointly modeled the target and the background for improving the tracker's discriminative power using subspace constraints and context fusion. Target information from multiple contexts and the surroundings are modeled using subspace constraints to reduce the impact of the boundary. The authors have developed the relationship between the target and its surroundings integrating spatial variations in contextual information [20]. Spatial variations prevent the tracker's drift due to external distractors and stabilize the CF's response for efficient tracking results. Multiple features are extracted and fused to obtain the unified feature response along with context information to enhance the target appearance. The peak-to-sidelobe ratio is to identify the highest feature response. Similarly, the authors have utilized multiple features in the tracker's appearance model in the CF framework to prevent the tracker's drift due to the limitation of single feature [23]. Group features along with context information are fused by the weighted sum of the normalized response maps. The tracker's adaptability to different sizes of target is improved by the adaptive padding method. Dynamic feature fusion along with the adaptive padding method improves tracker's robustness during tracking challenges. In summary, context-aware CF trackers integrate different context information and their surroundings to eliminate the negative impact of the boundary. Also, the tracker's potential is enhanced by incorporating multiple features in the tracker's appearance model.

10.2.2 Part-based correlation filter trackers

Part-based CF trackers learn target appearance based on the target's parts instead of tracking the whole target. Part-based CF trackers are based on the assumption that during complex occlusion and heavy background clutter, it is easier to track the target by its parts. Few parts are not affected and reliable enough to make robust tracking results. In this direction, the authors have integrated several local parts and global filters [4]. The global filters execute over the entire frame to locate the target when local parts fail. Hyper features are extracted from different layers of CNN to capture the fine details. Tracking is performed by collaborating the multiple parts and global filter along with adaptive selective filter update. Holistic and reliable local parts are extracted from the target in a kernelized CF [5]. Peak-to-sidelobe ratio computes the reliability of the local part to localize the target. When all the parts are unreliable, the sliding window approach generates several patches. These patches are tracked to obtain a reliable patch for the given part. Similarly, part-based tracking along with scale adaptation is proposed in a kernelized CF by [7]. Each part is trained by the individual classifier. Weighted coefficients are used for removing the abnormal matching points between the adjacent frames and handling the large-scale adaptations.

To handle the scale variations, the authors have estimated the target's scale from possible sets of scales in kernelized CF [17]. Multiple features are integrated for the effective discrimination of targets during the tracker's learning process. An adaptive set of filtering templates is used for capturing the target appearance variations. Li et al. [18] proposed target-aware CF, computing target likelihood to reduce the impact of background pixels on the foreground target pixels. The iterative optimization strategy is used for online filter learning. The authors encode the target's appearance using local and global filters jointly in a hybrid CF [19]. Local features localize target accurately and global feature prevents the tracker's drift during tracking challenges. The reliability of the tracker is measured using the peak-to-sidelobe ratio. In summary, part-based CF trackers are adaptive to various tracking challenges. Local parts and global parts are integrated to localize the target accurately. Parts reliability is computed to eliminate the negative impact of the affected parts to demonstrate superior tracking results.

10.2.3 Spatial regularization-based correlation filter trackers

CF tracker performance degrades due to the boundary effect of the target's surroundings. To address this, the spatial regularization term is incorporated during the tracker's filter training. For this, ℓ_1-norm spatial regularization is added to the tracker's objective to reduce the impact of noise in tracking results [9]. The target is localized by computing element-wise multiplication in the frequency domain. However, the ℓ_2-norm sparse regularization term is utilized to prevent the target's response map from being corrupted [21]. Sparse response regularization prevents the unexpected high peaks to corrupt the tracking output. This response also maintains the tracker's computational complexity. Zhu et al. [22] have proposed spatial attention in channel-specific discriminative CF trackers. Spatial sparsity is introduced to highlight the discriminative element in feature response during the filter learning stage. Spatial patterns are identified to remove the irrelevant and less significant channels. The authors have demonstrated spatial-temporal regularization in CF with Kalman filter to achieve tracking accuracy [8]. To prevent target loss due to sudden scale variations, the stride length control method is used. The impact of noise is diminished by estimating the target's motion state using a discrete-time Kalman estimator.

To address the ARC (aspect ratio change) challenge, Gaussian-like function labels are incorporated in CF during training [24]. The incremental update strategy is to update the samples with different aspect ratio adaptively and to prevent model degradation. However, the authors have proposed Gaussian scale mixture in Bayesian CF learning to eliminate the impact of boundary effects [25]. Structured Gaussian scale mixture utilizes spatial correlations among CG and jointly learns multipliers in a unified Bayesian framework. Also, the Gaussian scale mixture selects the most representation parameters for accurate target localization. In summary, the Gaussian function has shown superior tracking results in the CF framework by introducing joint learning and reliability. The impact of the boundary effect is minimized by the spatial regularization term in the parameter learning process during the tracker's learning.

10.3 DEEP CORRELATION FILTER-BASED TRACKERS

Tracking based on discriminative CF has shown tremendous performance during tracking variations. Table 10.2 shows the representative work exploiting deep features in a deep neural network-integrated with the CF framework. Deep features extract high-level and low-level discriminative features suitable for efficient tracking. The strength of CF and deep features are collaborated to propose a stable and reliable tracker. In this direction, the authors have utilized semantic and spatial convolution features to estimate the target state using scale CF [12]. Peak-to-sidelobe ratio is introduced in the update stage to measure the variations in the tracker's feature response. Spatial and semantic features are extracted from feature maps of *Conv1–2*, *Conv2–2*, *Conv3–4*, *Conv4–4*, and *Conv5–4* layers of a pre-trained VGG16-Net. Similarly, Zhang et al. [13] have extracted deep features from conv4-1, *conv4-2*, *conv4-3*, *conv5-1*, *conv5-2*, and *conv5-3* convolutional layers of VGG16-Net. Random projection matrix is generated using spectral graph theory to reduce the deep feature's dimension. This will improve the processing speed without sacrificing the trackers' accuracy. In Han et al. [26], the authors have coupled convolutional discriminative Fisher analysis into the correlation response of CF for the tracker's fine-tuning during tracking variations. The Fisher layer generates scene-specific features whereas the correlation layer models correlation response between adjacent frames for target appearance. Zheng et al. [27] have extracted multi-level features using VGG Net and explored the interdependencies between different levels of the feature. CF along with ADMM optimization is explored to train the network and model the network parameters. Adaptive update strategy ensures the tracker's effective update during environmental variations.

To address the limitations of the deep features-based tracker, CF, and PF are integrated by Mozhdehi and Medeiros [29]. Particles are provided as input to a CNN and correlation maps are generated using CF. PF localize the

Table 10.2 Description of representative work under Deep CF-based-tracking framework

SN	*References*	*Year*	*DL algorithm*	*Feature*	*Update strategy*	*Summary*
1.	Han et al. [26]	2018	CNN	Deep features	Iterative online update	Coupling of the outcome of the Fisher discriminative layer and CF layer to fine-tune the model to prevent drift.
2.	Zheng et al. [27]	2020	VGG Network	Deep features	–	Multi-level features with multitask learning to explore the importance of each feature.
3.	Zhang et al. [12]	2020	CNN	Spatial and semantic features	Online update	Multi-scale pyramid CF to extract the spatial and semantic features for robust tracking results.
4.	Zhang et al. [13]	2021	CNN	Deep features	Weighted update	Dimensionality reduction with random projection to minimize the tracker's computations.
5.	Liu and Liu [10]	2021	Siamese network	Semantic and spatial features	Weighted online update	Channel attentional mechanism to filter out noise from the frame and to reduce computations by extracting valuable features.
6.	Tian et al. [11]	2021	Siamese network	Deep features	Online update	Fuse multi-layer features to generate unified features to improve discriminative ability.
7.	Yuan et al. [28]	2021	Siamese network	Pre-trained feature extraction network	Linear model update	Multi-cycle consistency with deep feature extraction network to avoid manual sample extraction.
8.	Mozhdehia and Medeiros [29]	2022	Deep CNN	Deep features	Weighted update based on the previous and present state	Particle likelihood update after the iteration using K-means clustering with the consistent posterior distribution.
9.	Ruan et al. [30]	2022	Basic convolutional network	Deep features from the *conv2* and *conv5* layer	Linear weighted update	Spatial channel attention network, and channel-aware attention network for end-to-end feature learning for robust tracking.
10.	Nai et al. [31]	2023	CNN	HOG, color names, intensity, and CNN features	Lagrangian Multiplier-based Update	Discriminative spatial features selection to distinguish the target from the background.
11.	Zhou et al. [32]	2023	CNN	HOG and deep features from VGG Net	Linear update	Sample reliability-based active tracker's update for robust tracking.

The work is arranged in the ascending order of their published year.

target to the correct position by shifting them whereas CF determines the target's new position by analyzing the displacement between particles. Each particle extracts hierarchical CNN features and correlation response map to refine particle position. The authors have proposed a unified network for end-to-end learning-based on a target-background awareness model, spatial channel attention network, and a distractor-aware filter [30]. Tracker's representation learning is enhanced by generating attentive weights for channel attention and spatial attention network. Distractor-aware filters suppress background noise for a better response map for final target state estimation. Nai et al. [31] distinguish the target from the background by dynamically selecting the effective target's spatial features. Regularization term is considered in tracker's objective model for multitask sparse learning. Complementary handcrafted features and CNN features are learned simultaneously in CF. In Zhou et al. [32], the authors modeled two filters: the scale filter and the position filter. The scale filter is trained using handcrafted HOG whereas the position filter is equipped with shallow features from VGG-M and semantic features from VGG-16. Strong feature representation in CF framework to enhance tracker's robustness.

Siamese networks generate discriminative features for fast and accurate tracking results. In this direction, Liu et al. [10] have adaptively fused low-level spatial features with high-level semantic features guided by residual semantic embedding modules. The channel attention mechanism is incorporated to select valuable features by filtering out the background noise. Target features are learned offline for end-to-end tracker training. The authors have addressed the imbalance between non-semantic and semantic distractions for the tracker's efficient learning [11]. The flow-based tracking method is incorporated to handle occlusion. In Yuan et al. [28], the authors have extracted deep features from a pre-trained network and tracked the target using forward tracking. The similarity dropout strategy is to identify the low-quality samples and trajectory consistency loss to improve the training loss function. In summary, deep features-based CF trackers employ a pre-trained network not only to reduce the requirement of manual sampling but also to enhance the processing speed and robustness of the tracker.

10.4 FUSION-BASED CORRELATION FILTER TRACKERS

CF trackers provide robust tracking results by adopting strong feature extraction methods. CF trackers extract handcrafted features as well as deep features to capture the discriminative details of the target during environmental variations. Context-aware strategy and part-based methodology are adopted to prevent tracking failures. Regularization term and attention networks are also integrated for an efficient tracker's appearance model. But these techniques are not able to capture the interrelationship between the various features. The relationship between various features is essential to enhance the tracker's discriminative power during tracking challenges and hence, to improve tracking outcomes. The relationship between complementary features is upgraded by fusing them adaptively. Depending on the type of extracted feature, fusion-based CF trackers are categorized either as single modal-based CF trackers or multi modal-based CF trackers. Table 10.3 illustrates the salient features of the various proposed studies under fusion-based CF algorithms.

10.4.1 Single-model-based correlation filter trackers

Generally, trackers under this category extract multiple features from vision sensors. In this direction, Jain et al. [14] have extracted handcrafted features and deep features to capture the target's discriminative details. Channel graph regularization ensures that equal weights are assigned to similar feature channels to discriminate reliable features with rich target information. However, the authors have utilized spatial aware adaptive weight strategy in the CF framework [16]. Target pixels are assigned higher weights in comparison to background pixels to precisely localize the target. The update strategy prevents the tracker's contamination with corrupted background pixels during tedious tracking challenges. In Li and Yang [33], the authors have adopted a variable scale template strategy to mitigate the impact of background during target tracking. Complementary handcrafted features are extracted to handle the target's appearance variations. The high dimension of the feature vector is handled by processing feature-weighted responses in multiple classifiers. The quality and reliability of the template are measured by peak-to-sidelobe ratio.

The tracking performance is drastically improved by fusing the deep features with handcrafted features by handling the abrupt variations in target appearance. For this, the authors have extracted dynamic appearance variations by collaborating complementary information between adjacent frames [37]. Shu et al. [38] have adaptively fused multiple features using the weighted coefficient method. This fusion will relocate the target along with the occlusion handling mechanism when occlusion occurs. The target scale is predicted from the scale pool by classification search. The authors have proposed multi-expert feature representation using handcrafted and deep features [39]. Game theory is used to identify reliable experts to address tracking challenges. Interaction between the experts is modeled to identify high-score experts and adaptive updates to prevent tracker contamination. In Zhang et al. [9], the authors have fused feature maps from handcrafted features and pre-trained CNN deep features using the optimal soft mask. The consistency constraint term is introduced in the learning formula and soft mask to highlight the target information from the background. In summary, fusion-based CF trackers highlight the target's information from the background by suppressing the unimportant features and highlighting the relevant information. The tracker is updated adaptively to prevent contamination by unreliable samples.

Table 10.3 Description of representative work under fusion-based CF-tracking framework

SN	*References*	*Year*	*Sensor*	*Feature extraction*	*Fusion strategy*	*Summary*
1.	Li and Yang [33]	2019	Vision	Gabor energy, HOG, and color naming	Weighted feature-level fusion	To deal with the dimensionality issue, bi-features fused from the three classifiers.
2.	Zhai et al. [34]	2018	Vision and depth	HOG and depth	—	Particle weight updated and optimized for accurate prediction of the target position.
3.	Kuai et al. [15]	2019	Vision and depth	Color and depth	—	Segmentation map to highlight the target area for efficient tracking, occlusion detection, and occlusion handling.
4.	Yu et al. [35]	2019	Vision, thermal and IR	Gradient, intensity, and motion	Adaptive peak to correlation energy-based feature fusion	Effective multi-modal target representation and robust feature fusion to improve tracking accuracy.
5.	Luo et al. [36]	2019	Vision, thermal and IR	HOG, color names and HOG	Adaptive weighted fusion-based on a denoising scheme	Predict the target's position from two different modules and fuse them to final localize the target.
6.	Tang and Ling [16]	2019	Vision	HOG and color names	Feature likelihood map fusion	High confidence-based update and training of the tracking to prevent drift during occlusion.
7.	Zhu et al. [37]	2021	Vision	HOG, color names, and deep features	Weighted online fusion	Complementary feature fusion between the adjacent frames to extract dynamic information.
8.	Shu et al. [38]	2021	Vision	HOG, color names, edge and ULBP	Weighted fusion	Adaptive fusion of features along with occlusion judgment mechanism and scale filter.
9.	Ma et al. [39]	2022	Vision	HOG, color names, and deep features	Game theory-based decision level fusion	High confidence score experts select complementary information to improve tracking performance.
10.	Jain et al. [14]	2022	Vision	HOG and deep features from Norm1 of VGG-M, and Conv4-3 of VGG-16	Channel-aware graph regularization	Channel regularization and attention mechanism to reduce the impact of affected features on the tracker's performance.
11.	Zhang et al. [40]	2022	Vision	HOG and deep features from VGG-M and VGG-16	Single-scale feature fusion	Optimal translation and scale filter to adapt the trackers to tracking variations.
12.	Nai et al. [41]	2022	Vision	Color names, HOG, and deep features	Lagrangian Multiplier-based Update	Selective and representative features from the channels to improve the tracker's efficiency.

The works are arranged in ascending order of year of publication.

10.4.2 Multi-modal-based correlation filter trackers

With the availability of the target's depth and thermal information, CF-based trackers have gained popularity and provide excellent results. Depth information is efficient to handle occlusion while the thermal profile is suitable to track the target during the night and heavy background clutters. This information from specialized sensors is integrated with vision data in multi modal-based CF trackers. In this direction, Zhai et al. [34] have integrated HOG and depth in the CF framework for predicting the target's state. Depth data is used for occlusion handling and adaptive update to track the target during no occlusion. Scale variations mechanism to handle the varying target size for optimized tracking results. On the other hand, vision, thermal, and IR information is collaborated in continuous CFs [35]. Average peak correlation energy is utilized for adaptive feature fusion. Tracker overfitting is avoided by factorized convolution operator. In Luo et al. [36], the authors have extracted multi-modal information from the target's response map of vision and thermal profile. The target is tracked using two modules: histogram tracking and CF-tracking. Adaptive fusion strategy based on a weighting scheme ensures the denoising of the tracked frame by integrating the response from two modules. In summary, multi modal-based CF trackers demonstrate excellent performance by handling the target's tedious deformations and variations.

10.5 DISCUSSION ON CORRELATION FILTER-BASED TRACKERS

CF-based trackers locate the target in each frame by determining the highest response map in the target's search space. To maintain the computational processing of the trackers, CF computes the response map in the frequency domain [3, 17]. CF-based trackers provide stable tracking results in tedious tracking scenarios. But there are certain limitations associated with CF-based trackers that need to be addressed for efficient tracking solutions.

The CF-based trackers tend to drift when the target's sample space is contaminated with noise. This noise and interference affect the tracking results by shifting the location of the maximum feature response. To address this, handcrafted features are integrated with CF framework [23, 24]. Also, an adaptive weighting-based update strategy is employed to prevent the false update of the tracker [10, 29].

Deep learning-based trackers suffer from the requirement of large manually annotated samples. Extracting annotated samples is not only time-consuming but also computationally complex. This situation is handled by incorporating the CF in the deep learning networks [2, 28]. CF trackers based on an attention mechanism reduce the requirement of large training samples and also extracts the features rich in semantic information [10, 11, 29].

Also, DL-based trackers extract features from different layers of a complex network to generate robust features [42]. However, feature extraction from a complex and high dimension network is not only time-consuming, but is also impacted by the tracker's speed [43]. Hence, CF is incorporated with deep neural network to reduce the requirements of online training. CF-based trackers extract the high-level features from pre-trained networks such as VGG-16, VGG-M, and ResNet for the tracker's real-time processing speed [9, 24, 25].

10.6 SUMMARY

In this chapter, we have elaborated on the recent categories of CF-based trackers. In conventional CF-based trackers, context-aware, part-based, and spatial regularization-based CF trackers are explored. Context-aware CF trackers determine the target's contextual information to design the robust appearance model. Part-based CF trackers model the target region into multiple parts. The target's parts are analyzed to determine what is reliable during complex deformation and occlusion. The spatial regularization term is incorporated into the tracker's objective function to minimize the boundary effect. The boundary effect prevents the discrimination of the target from the background during tracking challenges.

The benefits of deep neural networks are integrated with CF trackers. It has been investigated that handcrafted features contain less discriminative spatial and semantic information in comparison to deep features. Hence, deep features are integrated with the CF framework to develop a robust feature extraction model. In addition, CF reduces the requirement of large training samples by incorporating an attention mechanism. Also, hierarchical features are extracted from a pre-trained network to ensure the tracker's real-time processing speed.

Fusion-based CF trackers capture the interrelationship between the various complementary features such as handcrafted features, deep features, and specialized features. The adaptive fusion strategy boosts the relevant features and suppresses the irrelevant features to ensure robust tracking results. Depth information is incorporated to handle the heavy occlusion and the thermal profile maintains CF tracker's performance during complex backgrounds and varying lighting conditions.

REFERENCES

1. Du, S. and S. Wang, An overview of correlation-filter-based object tracking. *IEEE Transactions on Computational Social Systems*, 2022. 9(1): pp. 18–31.
2. Liu, P., C. Liu, W. Zhao, and X. Tang, Multi-level context-adaptive correlation tracking. *Pattern Recognition*, 2019. 87: pp. 216–225.

3. Xu, J., C. Cai, J. Ning, and Y. Li, Robust correlation filter tracking via context fusion and subspace constraint. *Journal of Visual Communication and Image Representation*, 2019. **62**: pp. 182–192.
4. Barcellos, P. and J. Scharcanski, Part-based object tracking using multiple adaptive correlation filters. *IEEE Transactions on Instrumentation and Measurement*, 2021. **70**: pp. 1–10.
5. Li, C. and B. Yang, Correlation filter-based visual tracking via holistic and reliable local parts. *Journal of Electronic Imaging*, 2019. **28**(1): pp. 013039–013039.
6. Li, C., X. Liu, X. Su, and B. Zhang, Robust kernelized correlation filter with scale adaptation for real-time single object tracking. *Journal of Real-Time Image Processing*, 2018. **15**(3): pp. 583–596.
7. Lian, G.-Y., A novel real-time object tracking based on kernelized correlation filter with self-adaptive scale computation in combination with color attribution. *Journal of Ambient Intelligence and Humanized Computing*, 2020: pp. 1–9.
8. Feng, S., K. Hu, E. Fan, L. Zhao, and C. Wu, Kalman filter for spatial-temporal regularized correlation filters. *IEEE Transactions on Image Processing*, 2021. **30**: pp. 3263–3278.
9. Zhang, J., Y. He, and S. Wang, Learning adaptive sparse spatially-regularized correlation filters for visual tracking. *IEEE Signal Processing Letters*, 2023. **30**: pp. 11–15.
10. Liu, G. and G. Liu, End-to-end correlation tracking with enhanced multi-level feature fusion. *IEEE Access*, 2021. **9**: pp. 128827–128840.
11. Tian, L., P. Huang, Z. Lin, and T. Lv, DCFNet++: More advanced correlation filters network for real-time object tracking. *IEEE Sensors Journal*, 2021. **21**(10): pp. 11329–11338.
12. Zhang, J., X. Jin, J. Sun, J. Wang, and A.K. Sangaiah, Spatial and semantic convolutional features for robust visual object tracking. *Multimedia Tools and Applications*, 2020. **79**(21): pp. 15095–15115.
13. Zhang, M., L. Xu, J. Xiong, and X. Zhang, Correlation filter via random-projection based CNNs features combination for visual tracking. *Journal of Visual Communication and Image Representation*, 2021. **77**: p. 103082.
14. Jain, M., A. Tyagi, A.V. Subramanyam, S. Denman, S. Sridharan, and C. Fookes, Channel graph regularized correlation filters for visual object tracking. *IEEE Transactions on Circuits and Systems for Video Technology*, 2022. **32**(2): pp. 715–729.
15. Kuai, Y., G. Wen, D. Li, and J. Xiao, Target-aware correlation filter tracking in RGBD videos. *IEEE Sensors Journal*, 2019. **19**(20): pp. 9522–9531.
16. Tang, F. and Q. Ling, Spatial-aware correlation filters with adaptive weight maps for visual tracking. *Neurocomputing*, 2019. **358**: pp. 369–384.
17. Bai, B., B. Zhong, G. Ouyang, P. Wang, X. Liu, Z. Chen, and C. Wang, Kernel correlation filters for visual tracking with adaptive fusion of heterogeneous cues. *Neurocomputing*, 2018. **286**: pp. 109–120.
18. Li, D., G. Wen, Y. Kuai, J. Xiao, and F. Porikli, Learning target-aware correlation filters for visual tracking. *Journal of Visual Communication and Image Representation*, 2019. **58**: pp. 149–159.
19. Wang, Y., X. Luo, L. Ding, J. Wu, and S. Fu, Robust visual tracking via a hybrid correlation filter. *Multimedia Tools and Applications*, 2019. **78**(22): pp. 31633–31648.

20. Elayaperumal, D. and Y.H. Joo, Robust visual object tracking using context-based spatial variation via multi-feature fusion. *Information Sciences*, 2021. **577**: pp. 467–482.
21. Zhang, W., L. Jiao, Y. Li, and J. Liu, Sparse learning-based correlation filter for robust tracking. *IEEE Transactions on Image Processing*, 2021. **30**: pp. 878–891.
22. Zhu, X.-F., X.-J. Wu, T. Xu, Z.-H. Feng, and J. Kittler, Robust visual object tracking via adaptive attribute-aware discriminative correlation filters. *IEEE Transactions on Multimedia*, 2022. **24**: pp. 301–312.
23. Feng, Z., L. Yan, Y. Xia, and B. Xiao, An adaptive padding correlation filter with group feature fusion for robust visual tracking. *IEEE/CAA Journal of Automatica Sinica*, 2022. **9**(10): pp. 1845–1860.
24. Chen, J., T. Xu, B. Huang, Y. Wang, and J. Li, ARTracker: Compute a more accurate and robust correlation filter for UAV tracking. *IEEE Geoscience and Remote Sensing Letters*, 2022. **19**: pp. 1–5.
25. Cao, Y., G. Shi, T. Zhang, W. Dong, J. Wu, X. Xie, and X. Li, Bayesian correlation filter learning with Gaussian scale mixture model for visual tracking. *IEEE Transactions on Circuits and Systems for Video Technology*, 2022. **32**(5): pp. 3085–3098.
26. Han, Z., P. Wang, and Q. Ye, Adaptive discriminative deep correlation filter for visual object tracking. *IEEE Transactions on Circuits and Systems for Video Technology*, 2018. **30**(1): pp. 155–166.
27. Zheng, Y., X. Liu, X. Cheng, K. Zhang, Y. Wu, and S. Chen, Multitask deep dual correlation filters for visual tracking. *IEEE Transactions on Image Processing*, 2020. **29**: pp. 9614–9626.
28. Yuan, D., X. Chang, P.-Y. Huang, Q. Liu, and Z. He, Self-supervised deep correlation tracking. *IEEE Transactions on Image Processing*, 2021. **30**: pp. 976–985.
29. Mozhdehi, R.J. and H. Medeiros, Deep convolutional correlation iterative particle filter for visual tracking. *Computer Vision and Image Understanding*, 2022. **222**: p. 103479.
30. Ruan, W., M. Ye, Y. Wu, W. Liu, J. Chen, C. Liang, ... C.-W. Lin, Ticnet: A target-insight correlation network for object tracking. *IEEE Transactions on Cybernetics*, 2022. **52**(11): pp. 12150–12162.
31. Nai, K., Z. Li, Y. Gan, and Q. Wang, Robust visual tracking via multitask sparse correlation filters learning. *IEEE Transactions on Neural Networks and Learning Systems*, 2023.
32. Zhou, L., J. Li, B. Lei, W. Li, and J. Leng, Correlation filter tracker with sample-reliability awareness and self-guided update. *IEEE Transactions on Circuits and Systems for Video Technology*, 2023. **33**(1): pp. 118–131.
33. Li, C. and B. Yang, CFGVF: An improved correlation filters based visual tracking algorithm. *Optik*, 2019. **192**: p. 162930.
34. Zhai, Y., P. Song, Z. Mou, X. Chen, and X. Liu, Occlusion-aware correlation particle filter target tracking based on RGBD data. *IEEE Access*, 2018. **6**: pp. 50752–50764.
35. Yu, T., B. Mo, F. Liu, H. Qi, and Y. Liu, Robust thermal infrared object tracking with continuous correlation filters and adaptive feature fusion. *Infrared Physics & Technology*, 2019. **98**: pp. 69–81.

36. Luo, C., B. Sun, K. Yang, T. Lu, and W.-C. Yeh, Thermal infrared and visible sequences fusion tracking based on a hybrid tracking framework with adaptive weighting scheme. *Infrared Physics & Technology*, 2019. **99**: pp. 265–276.
37. Zhu, X.-F., X.-J. Wu, T. Xu, Z.-H. Feng, and J. Kittler, Complementary discriminative correlation filters based on collaborative representation for visual object tracking. *IEEE Transactions on Circuits and Systems for Video Technology*, 2021. **31**(2): pp. 557–568.
38. Shu, Q., H. Lai, L. Wang, and Z. Jia, Multi-feature fusion target re-location tracking based on correlation filters. *IEEE Access*, 2021. **9**: pp. 28954–28964.
39. Ma, S., Z. Zhao, Z. Hou, L. Zhang, X. Yang, and L. Pu, Correlation filters based on multi-expert and game theory for visual object tracking. *IEEE Transactions on Instrumentation and Measurement*, 2022. **71**: pp. 1–14.
40. Zhang, K., W. Wang, J. Wang, Q. Wang, and X. Li, Learning adaptive target-and-surrounding soft mask for correlation filter based visual tracking. *IEEE Transactions on Circuits and Systems for Video Technology*, 2022. **32**(6): pp. 3708–3721.
41. Nai, K., Z. Li, and H. Wang, Learning channel-aware correlation filters for robust object tracking. *IEEE Transactions on Circuits and Systems for Video Technology*, 2022. **32**(11): pp. 7843–7857.
42. Kumar, A., G. S. Walia, & K. Sharma, Recent trends in multicue based visual tracking: A review. *Expert Systems with Applications*, 2020. **162**: p. 113711.
43. Kumar, A., R. Jain, V. A. Devi, & A. Nayyar, (Eds.). *Object Tracking Technology: Trends, Challenges, Impact, and Applications*, 2023. Springer.

Chapter 11

Future prospects of visual tracking

Application-specific analysis

11.1 INTRODUCTION

DL-based visual tracking extracts deep features such as deep color, deep optical flow, and deep LBP from CNN [1, 2]. Hierarchical features from different convolutional layers are also extracted for better feature representations [3, 4]. Deep features and hierarchical features are more discriminative and richer in the target's spatial and semantic information. However, the complex and high dimension of the deep neural network increases the computational complexity of the network. In addition, it slows the processing speed of the tracker and makes it unsuitable for real-time tracking.

To address the deep and complex architecture of the neural network, pruning techniques are explored in deep neural networks [5, 6]. The large number of parameters, weight training, and intermediate layers increase the complexity of the trackers. High computing and training requirements of deep neural networks restrict their applicability to resource-constrained devices. It has been observed that the Siamese network and CNN demand extensive computing and henceforth, reduce the tracker's speed. Pruning techniques compress deep neural network architecture by identifying redundant weights, filters, or layers [7–10]. These techniques improve network efficiency by reducing the computations with negligible accuracy loss. The network compression ratio will determine the potential of the network in terms of accuracy loss and the number of pruned parameters.

The black box nature of deep neural networks does not identify the feature importance for efficient tracking solutions [11]. Deep neural networks are not interpretable in terms of feature visualization for lost targets and re-localization of the target trajectory. Feature relevance is essential for the generalizability of algorithms to diverse domains [12]. Video surveillance systems can track specific targets only for which they are trained [13]. Also, the processing is not real-time due to the high dimension of the neural network. Deep neural networks should be explainable, interpretable, and generalizable to expand their applicability to diverse domains such as crowd monitoring, surveillance, and many more.

DOI: 10.1201/9781003456322-11

Recently, visual tracking is widely explored in various tracking domains that include pedestrian tracking [14], gaze tracking [15], 3-D tracking [16], and autonomous vehicle path tracking [17]. Pedestrian tracking keeps track of the activities in dense areas where CCTV cameras are installed. This helps in crowd monitoring to prevent crime and ensure security using lightweight Siamese networks [18]. 3-D tracking obtains the target's context key points to localize the target. Context key points are learned by spatio-temporal relationship between the coordinates on the target [19]. Autonomous vehicle surveillance and tracking are necessary to prevent fatal casualties on the road. Self-driving cars are at high risk of crashes and accidents, hence accurate real-time systems for tracking are necessary so that proper actions can be taken on time. In summary, visual tracking has penetrated many real-time activities to provide a secure and safe environment for movement. From pedestrian tracking to autonomous vehicle tracking, visual tracking is emerging as an interactive field by fulfilling the sustainable goal of smart city development.

11.2 PRUNING FOR DEEP NEURAL ARCHITECTURE

Deep neural networks have shown remarkable performance in computer vision. In visual tracking, DL-based trackers have shown more stable and accurate tracking results in comparison to trackers based on conventional approaches. However, high hardware, computing, and storage requirements limit their deployment to handheld devices which have low hardware specifications. To address this issue, pruning techniques are explored to minimize the network size and complexity. The pruning technique determines redundant weights, filters, networks, and channels in a neural network. These redundant parameters play no significant role in algorithm computations and their exclusion from the network does not impact the model's accuracy. The next section will detail the various types of pruning techniques.

11.2.1 Types of pruning network

Before pruning a network, the granularity of the network needs to be identified in terms of which parameters are to be pruned. Based on the granularity of the pruning unit, pruning techniques in the deep neural network are broadly categorized as unstructured pruning and structured pruning techniques. Figure 11.1 illustrates the various pruning techniques. Unstructured pruning techniques have fine granularity and tend to identify the redundant smallest network unit such as neuron weight or kernel [9, 28]. Whereas structured pruning techniques have a coarse level granularity that is based on filter, channel or layer [5, 7, 22, 24]. Table 11.1 shows the recent works under various pruning techniques. The main objective of the pruning

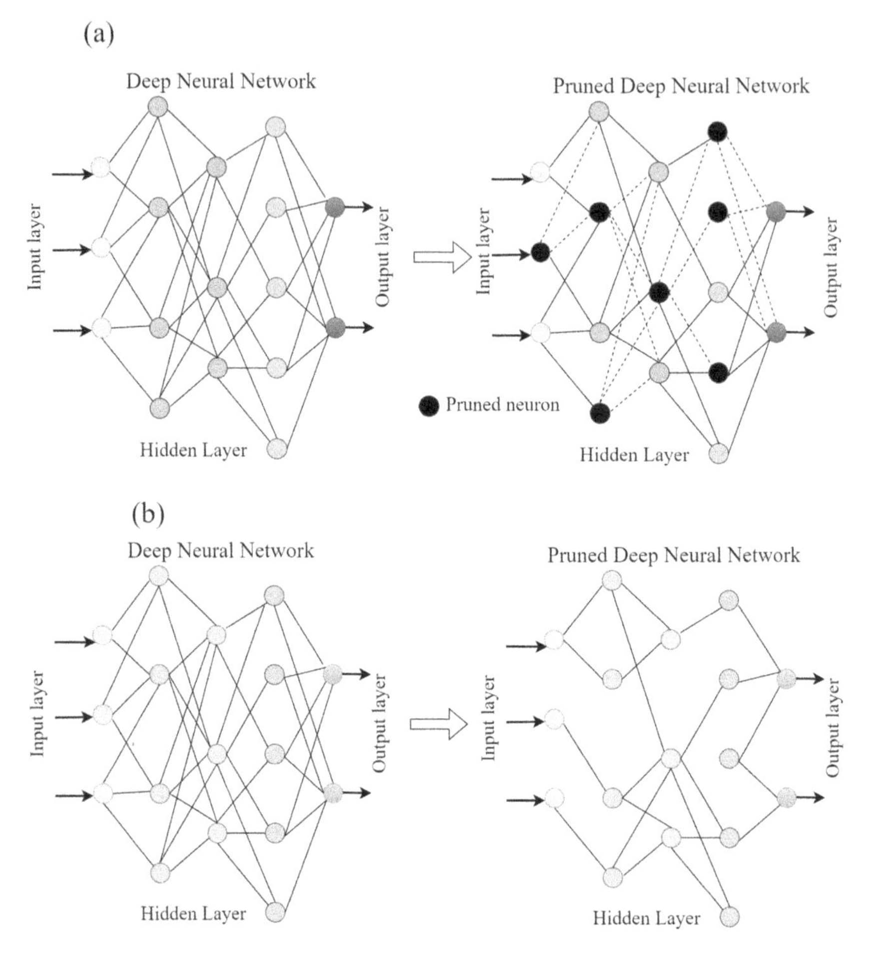

Figure 11.1 Types of pruning techniques (a) Structured pruning (b) Unstructured pruning.

technique is to reduce the complexity of the network without sacrificing much accuracy.

Unstructured pruning as weight pruning in the deep neural network is explored by ref. [9, 28]. In Wu et al. [9], the authors have proposed different levels of pruning for different layers of deep neural networks. Differential evolutionary algorithms determine the pruning sensitivity for each layer by iterating weight pruning. Optimization is applied for the sensitivity for each layer to obtain optimal weights. The authors have reduced the size of the network by utilizing state, action, and reward as key points in the pruning context [28]. State determines as subspaces for layer index and layer density. Action space deals with layer sparsity to determine layers with the smallest weights. Reward stage computes the loss and change in loss due to action

Table 11.1 Representative work under various pruning architectures

SN	*References*	*Year*	*Type of pruning*	*DL networks*	*Methodology*	*Summary*
1.	Wang et al. [5]	2021	Filter pruning	VGG16, and ResNet50	Granularity-based feature entropy map	Define feature map importance to identify the redundant filter weights.
2.	Zhang and Wang [20]	2022	Filter pruning	VGG16, DenseNet40, ResNet18 and ResNet50	Saliency-based distance between filters strides	Compute the similarity distance along with the saliency map to determine important filter weights.
3.	Wang et al.[21]	2020	Network pruning	VGG-16/19, and ResNet-20/50/164	Sparse learning and genetic algorithm	Identify important weights based on dynamic adjustable evaluation factors with optimization.
4.	Chen et al. [7]	2021	Channel pruning	VGG-11/16/19, ResNet-20/32, and ResNet-50/56/110	Sparse regularization and channel selection	Combine weights from two layers of the deep neural network to identify important channels.
5.	Lian et al. [22]	2021	Filter pruning	ResNet 18 and ResNet56	Sparsity search and model training	Sparsity search with an evolutionary algorithm to identify relevant filters for pruning.
6.	Zhu and Pei [23]	2022	Kernel pruning	VGG16, ResNet-18/34/50/101, ResNet-20/32/56/110 and MobileNet V2	Saliency mapping	Dual port saliency mapping to identify the kernel pruning.
7.	Chen et al. [24]	2021	Channel pruning	ResNet-18/50 and MobileNet V2	Sequential interval estimation	Utilize gradient descent in each group to search for the optimal substructure as per the distribution of channels.
8.	Ma et al. [25]	2022	Filter pruning	AlexNet, VGG-16 and ResNet18	Mask-aware convolutional computations	Mask-based filter pruning to identify redundant filters.
9.	Wang et al. [26]	2022	Weight pruning	Graph neural network	Graph edge pruning	Saliency matrix to measure the graph edge importance and prune edges with negative scores.
10.	Tian et al. [27]	2021	Filter pruning	VGG19, ResNet-50/164 and ResNet-56/110	One stage pruning with gradient estimator	The sparse structure and weight optimization while training the model.
11.	Wu et al. [9]	2021	Weight pruning	LeNet-300/100, LeNet5, AlexNet and VGG16	Differential evolution	Recovery strategy to add the removed connections again during network fine-tuning.
12.	Camci et al. [28]	2022	Weight pruning	ResNet-32/56/50 and MobileNet V1	Q-Learning	Greedy action, reward, and state-based update and pruning strategy.

spaces. On the other hand, Zhu and Pei [23] have proposed kernel pruning using the single port and dual port salient mapping channels. Salient convolutional kernel's average mapping ability measurement index is computed to switch between single port and dual port kernel salient mapping channels. The above weight pruning techniques efficiently compress a heavy-weight deep neural network with minimal deviation from original network outcomes.

Filter or channel pruning is one of the most effective types of structured pruning techniques. In Zhang and Wang [20], the authors have explored filter similarity by determining distance weights between filters. Multiple distance metrics are calculated to determine the redundant filters based on the similarity scores. Lian et al. [22] have utilized an evolutionary algorithm based on discrete particle swarm optimization to select the optimal number of filters to be pruned. Initially, sparsity search along with training is performed in a pre-trained network to identify the filters that can be pruned. However, the authors have considered Pearson correlation to compute the correlation between the filters to identify them either as unimportant or important [6]. Penalty terms based on parameter quantity and computational cost are considered to determine the unimportant filters. The importance of the filter is determined in accordance with the entropy of feature maps [5]. Feature maps with low entropy values are considered to contain less or no information, hence the corresponding filter is considered redundant and to be removed from the network.

Apart from filter pruning, channels and network are also pruned to reduce the complexity of the deep neural network. In this direction, Chen et al. [7] have proposed collaborative channel pruning to determine importance. The weights of the convolutional layers and the scale factor of batch normalization layers are combined to compute the channel importance. Channels that contain no information are removed from the network to reduce the computations. However, the authors have proposed sparse learning and genetic algorithm to identify the redundant network channels [21]. $\ell_{1/2}$ regularization term is incorporated into model training to achieve channel-level sparsity. Genetic algorithms are used to identify the importance of the subnetwork obtained from the sparsity space. In summary, channel or network pruning techniques reduce the network size by evaluating the important parameters. Parameters contribution is calculated to evaluate them as redundant and henceforth, enhance network computational complexity by removing them.

11.2.2 Benefits of pruning

Recently, there has been an increasing trend in deep neural network-based tracking architecture. But high computing requirements restrict their deployment on handheld and portable devices. During the investigation, it has been observed that there are many redundant parameters in the network. Once identified, they can be removed from the network without any drop in accuracy. The pruning techniques offer several benefits not only in improving the computations, but also reducing storage costs. The details of the benefits are as follows:

1. The pruning strategy determines redundant weights, filters, networks, and channels based on certain criteria. The removal of these fundamentals from the deep network does not impact its performance.
2. Pruning techniques can reduce computational processing by minimizing the floating-point operations (FLOPs) during the iterative process of network training.
3. As the numbers of hyperparameters are reduced and the channels are compressed, pruning techniques speed up the overall network computations.
4. Pruning techniques ensure negligible or minimum sparsity loss for the computation of the network parameters to determine the output.

Apart from mentioned benefits, pruning techniques have certain limitations. The details are as follows:

1. After the removal of the redundant parameters, the deep neural network requires re-training of the network. Re-training is essential for fine-tuning the network to reduce accuracy loss. However, this step involves extra computations and so increases the processing time.
2. Identification of the redundant parameters is based on certain algorithms. If the pruning algorithms' complexity is high then it that aids in the increase in the system complexity and impacts the speed of the network.
3. The generalizability of the pruned networks is yet to be explored. The performance of pruned deep neural networks needs to be investigated for their applicability to unseen training data. The evaluation parameters such as accuracy loss, compression ratio, sparsity ratio, and others should be compared to prove the suitability of pruned network to real-time scenarios.

Hence the trade-off between the pruning algorithm complexity and compression ratio must be considered to ensure the overall speeding up of the deep neural network.

11.3 EXPLAINABLE AI

To build trust, transparency, and information flow visualization, it is essential to explain the prediction and outcomes of a deep neural network. In this direction, explainable AI (XAI) has come out with a solution to provide human interpretations to increase the adoption of DL model-based outcomes in real life [29]. Explaining DL and ML algorithms requires different steps depending on the algorithm's objective. Feature relevance, feature importance, fairness, accessibility, and privacy need to be understood to define the internal functionality of the network [30].

Explaining the evaluation in XAI is a tedious activity. It may be due to the black-box nature of the deep neural network. It is relatively difficult to define the quantity and quality of the deep neural network. The extraction of the output at different layers of the CNN is not understandable by the same set of explanations. The explanation related to the accuracy and clarity of the model architecture demands explanation so that the model predictions can be trusted. Human intervariability also needs to be considered so that the model explanation should be the same for all people. The model should be generic in its application and robust explanations must be provided to attribute the failure to any other scenario [31]. Also, local and global feature weighting methods are explained for the successful deployment of deep neural networks.

In deep neural networks, some transformation functions are applied to obtain the high-dimension discriminative features [32]. All these features are unified into a single robust feature vector suitable for making accurate predictions. The predictions are a joint effort of the local and global classifications computed from the network. The reliability of the outcomes needs to be determined to determine the reason which features contribute to predictions positively or negatively. Interpretation of the learned features by analyzing the network architecture is necessary for efficient learning of the system. For this, explainable deep learning models are categorized as attribution, and non-attribution-based methods [33]. Model compression and acceleration are necessary for the model's robustness and training stability.

XAI is the latest innovation to explain the model of the basis of semantic rules for classifier scores [34]. It has been suggested to represent the black box systems for the ML system outcomes. The methods for explaining the deep neural network such as Grad-CAM, SHAP, and SHAP-CAM are used for providing the attribution of each feature [35]. In addition, XAI on social media platforms using VADER, SHAP, and LIME are explored by Ref. [36]. The interpretability of the deep learning models is visualized by bar plots.

11.3.1 Importance of generalizability for deep neural networks

It has been observed that deep neural networks are data set-specific. In other words, deep neural networks perform well on data sets on which they are trained. However, their performance tends to deteriorate on other data sets. The DL models are not generalizable in their applicability. The design of deep neural should be independent and be able to address the tracking challenges. Discrimination based on data sets should be addressed either at pre-processing steps or post-processing steps [37]. Error in the data sets is embedded in the model outcome and hence produces biased results.

In Petch et al. [38], the authors have highlighted the need for interpretable, explainable, and generalizable DL models even at the cost of scarifying the accuracy. Classifier outcomes should be explainable to produce accurate results on unseen data sets. The classifier model rules are interpretable and understandable by humans. Hence the DL models should be equipped with adaptability, generalizability, and interpretability to analyze the variations in feature importance to provide reliable, robust, fair, and accurate structure.

11.4 APPLICATION-SPECIFIC VISUAL TRACKING

In this section, some recent applications of visual tracking are explored. Visual tracking has provided many recent solutions for monitoring crowds, security, surveillance, and navigation. The details for application-specific visual tracking are as follows.

11.4.1 Pedestrian tracking

Pedestrian tracking has attracted many researchers due to its potential applications for surveillance, tracking, and entertainment. In recent years, many crash-related deaths of pedestrians have been reported [39]. Hence it became imperative to track pedestrians to prevent fatal injuries. The authors have reviewed pedestrian detection and tracking methods based on conventional feature extraction techniques and DL-based techniques [39]. It has been inferred that accurate and reliable tracking results can be achieved by integrating traditional approaches with DL methods. Automatic pedestrian detection and tracking is the upcoming trend that needs to be followed for preventing pedestrian casualties.

In Sarcinelli et al. [14], the authors explore pedestrian detection and tracking to prevent crashes in self-driving cars. The method is based on Frenét frames which are integrated with a path generator as a decision-making system. The system analyses the pedestrian's behavior to decide whether to start, stop, or overtake the pedestrian. The accuracy of the system is tested on simulated videos as well as real-world scenarios.

In Bićanić et al. [40], the authors have utilized tracking by detection approach in a deep learning framework. The relationship between the kinematics and the appearance cues is established to track the pedestrian. Pre-trained R-CNN is exploited to provide efficient tracking results.

In Wong et al. [41], the authors have analyzed the walking behavior of pedestrians for tracking. High-level pedestrian attributes are integrated with multiple features extracted from the pedestrian. Similar measures and identity mechanisms are employed for enhancing the accuracy of the system. The authors have claimed that the model is generalizable in its applicability as it recognizes the pedestrian movement patterns neglecting the background disturbances.

In another direction, in ref. [42], the authors have integrated the thermal infrared characteristics of the object by considering the temporal and spatial information. For efficient results, the target's original appearance, current appearance, and the surrounding context are considered. The tracking algorithm is high-speed and suitable for real-time scenarios. In summary, pedestrian tracking algorithms should provide the results in real time so that necessary steps can be taken in case of emergency. Pedestrian tracking is essential to fulfilling the goal of smart cities having autonomous vehicles and self-driving cars.

11.4.2 Human activity tracking

Human activity tracking is widely explored to improve daily lives by analyzing behavioral activity [15]. It identifies human activities based on the outcomes produced by electronic sensors such as camera devices, CCTVs, and wearable devices to provide a wide range of applications [43]. In the healthcare domain, it provides health monitoring to elderly people to improve their lifestyles. Apart from this, it has diverse applications such as surveillance, physical activity tracking, entertainment, gaming, and many more [44].

The authors have utilized gaze ethograms to determine the behavioral activity of the target [15]. Six pre-trained public DL algorithms are exploited to generate the results on two diverse data sets. It has been observed that the incorporation of transfer learning in the DL model has reduced the computations and hence enhanced the performance of the model.

In Gupta et al. [43], the authors have reviewed the various AI applications for human activity recognition. The wide range of human activity is analyzed in the healthcare and non-healthcare domain. Data acquisition sensors and electronics devices along with their applications and exploited AI techniques are explored to provide strong future pointers. The recent use of technology such as Wi-Fi in human activity tracking is described as providing health monitoring facilities and related services to older people remotely. The rationalized design of human activity recognition algorithms with DL methodology along with transfer learning is explored to provide abnormal action prediction such as fall detection. Also, sensors and wearable devices are discussed to monitor vital medical statistics such as blood pressure, oxygen level, pulse rate monitoring, sleep monitoring, and many more.

In Zhang et al. [44], the authors have explored the role of wearable sensors in tracking human activities using DL methodologies. Wearable devices such as smart glasses, smart watches, smart gloves, and armbands are very popular to record human body movement and track day-to-day activities. These vital stats data recordings are used for tracking the intensity of physical activity to determine the risk of chronic devices such as obesity, cardiac vascular diseases, diabetes, and many more. Wearable devices also play a significant role in diagnosing critical alignments such as Parkinson's disease, and other mental abnormalities by collecting the patient's data based on their muscular activities and movements. These devices are equipped with the latest technology that provides real-time health monitoring and enables doctors to take suitable steps to prevent casualties. In summary, human activity monitoring has the potential opportunity to improve real-life applications. DL approaches are integrated with the data collected from the sensors and devices to benefit the user by predicting chronic health issue at an early stage.

11.4.3 Autonomous vehicle path tracking

Autonomous vehicle tracking is one of the latest domains to prevent road accident casualties by reducing human interventions [17]. It senses the surroundings and performs path tracking to reduce the potential risk of collision by automatic speed control, and navigating of the vehicle. The key benefits of autonomous vehicles are vital in reducing road accidents, enhancing productivity, and reducing travel costs. However, the requirement of numerous tasks, such as pedestrian detection, path tracking, object localization, traffic monitoring, and decision-making, makes autonomous vehicle tracking tedious and challenging [45]. In addition, the integration of captured data from various sensors demands quick processing so that real-time decisions can be taken. Identification, detection, monitoring, tracking, and decision-making should be integrated to address the uncertain driving conditions for safe and secure travel.

In Rokonuzzaman et al. [46], the authors have reviewed various path-tracking controllers for autonomous vehicles for effective and accurate tracking performance. Path tracking controllers are useful for path planning of autonomous vehicles considering the vehicle dynamics to control speed and other driving scenarios. Path-tracking controllers reduce the complexity of the autonomous vehicle by minimizing tracking errors and enhancing accuracy and robustness to prevent road accidents.

In Ruslan et al. [47], the authors have discussed various control strategies for path tracking of autonomous vehicles. Control strategies are the decision-making strategies to control the vehicle movement on a desired path. The path followed by path tracking controllers is to ensure that the vehicle should follow a pre-defined path determined by analyzing the surroundings. In summary, autonomous vehicle path tracking ensures effortless vehicle movement by computing the inputs from multiple sensors. The sensor data should be processed and integrated at a faster rate so that quick decisions can be taken to prevent casualties. Simplifying configuration and implementation of various tracking attributes without sacrificing speed, accuracy, and reliability are key challenges in autonomous vehicle path tracking. In addition, fine-tuning in accordance with changing road conditions, traffic density, and pedestrian localization are the most common characteristics that need to be introduced for better outcomes.

11.5 SUMMARY

In this chapter, we have discussed the pruning techniques to determine the redundant weights, filters, or channels to reduce the complex deep neural network architectures. The benefits and limitations of pruning techniques are elaborated to increase the processing speed of the DL modal. In addition, the salient features of pruning techniques are shown to explore the applicability of these techniques tracking domain. The pruning techniques compress the network size eliminating the redundant parameters with negligible loss of accuracy and significant improvement in speed.

Deep neural networks are not explainable in their outcome due to the black-box nature of the architecture. Network transparency and generalizability are the key factors that need to be ensured for trust and fairness in deep neural networks. DL networks also suffered from AI bias that introduced either due to data or algorithm exploited. Synthetic data used to increase the training data size is responsible for data bias in the system. However, the black box design of the DL model is the reason for algorithmic bias.

Tracking is explored widely in recent areas that include pedestrian tracking, human activity tracking, and autonomous vehicle tracking. Pedestrian tracking is the class of object tracking that tracks pedestrians to prevent casualties due to driverless cars and autonomous vehicle tracking. Human activity tracking records human physical statistics with the help of sensors and wearable devices. Sensors track human motion to detect the medical alignments in the person. Wearable devices are equipped with the latest technology providing real-time health monitoring records to the doctor. These devices are very helpful for elderly people for remote health monitoring and enhancing real-life experiences.

REFERENCES

1. Qian, X., L. Han, Y. Wang, and M. Ding, Deep learning assisted robust visual tracking with adaptive particle filtering. *Signal Processing: Image Communication*, 2018. **60**: pp. 183–192.
2. Xiao, J., R. Stolkin, M. Oussalah, and A. Leonardis, Continuously adaptive data fusion and model relearning for particle filter tracking with multiple features. *IEEE Sensors Journal*, 2016. **16**(8): pp. 2639–2649.
3. He, X. and Y. Sun, SiamBC: Context-related Siamese network for visual object tracking. *IEEE Access*, 2022. **10**: pp. 76998–77010.

4. Li, X., Q. Liu, N. Fan, Z. He, and H. Wang, Hierarchical spatial-aware Siamese network for thermal infrared object tracking. *Knowledge-Based Systems*, 2019. **166**: pp. 71–81.
5. Wang, J., T. Jiang, Z. Cui, and Z. Cao, Filter pruning with a feature map entropy importance criterion for convolution neural networks compressing. *Neurocomputing*, 2021. **461**: pp. 41–54.
6. Wang, W., Z. Yu, C. Fu, D. Cai, and X. He, COP: Customized correlation-based Filter level pruning method for deep CNN compression. *Neurocomputing*, 2021. **464**: pp. 533–545.
7. Chen, Y., X. Wen, Y. Zhang, and W. Shi, CCPrune: Collaborative channel pruning for learning compact convolutional networks. *Neurocomputing*, 2021. **451**: pp. 35–45.
8. Ma, M. and J. An, Combination of evidence with different weighting factors: A novel probabilistic-based dissimilarity measure approach. *Journal of Sensors*, 2015. **2015**.
9. Wu, T., X. Li, D. Zhou, N. Li, and J. Shi, Differential evolution-based layer-wise weight pruning for compressing deep neural networks. *Sensors*, 2021. **21**(3): p. 880.
10. Zhou, A., A. Yao, Y. Guo, L. Xu, and Y. Chen, Incremental network quantization: Towards lossless CNNs with low-precision weights. arXiv preprint arXiv:1702.03044, 2017.
11. Ntoutsi, E., P. Fafalios, U. Gadiraju, V. Iosifidis, W. Nejdl, M.E. Vidal, ... E. Krasanakis, Bias in data-driven artificial intelligence systems—An introductory survey. *Wiley Interdisciplinary Reviews: Data Mining and Knowledge Discovery*, 2020. **10**(3): p. e1356.
12. Himeur, Y., S. Al-Maadeed, H. Kheddar, N. Al-Maadeed, K. Abualsaud, A. Mohamed, and T. Khattab, Video surveillance using deep transfer learning and deep domain adaptation: Towards better generalization. *Engineering Applications of Artificial Intelligence*, 2023. **119**: p. 105698.
13. Kumar, A., G. S. Walia, & K. Sharma, Recent trends in multicue based visual tracking: A review. *Expert Systems with Applications*, 2020. **162**: p. 113711.
14. Sarcinelli, R., R. Guidolini, V.B. Cardoso, T.M. Paixão, R.F. Berriel, P. Azevedo, ... T. Oliveira-Santos, Handling pedestrians in self-driving cars using image tracking and alternative path generation with Frenét frames. *Computers & Graphics*, 2019. **84**: pp. 173–184.
15. de Lope, J. and M. Graña, Deep transfer learning-based gaze tracking for behavioral activity recognition. *Neurocomputing*, 2022. **500**: pp. 518–527.
16. Nguyen, U. and C. Heipke, 3d pedestrian tracking using local structure constraints. *ISPRS Journal of Photogrammetry and Remote Sensing*, 2020. **166**: pp. 347–358.
17. Kumar, A., R. Jain, V. A. Devi, & A. Nayyar, (Eds.). *Object Tracking Technology: Trends, Challenges, Impact, and Applications*, 2023. Springer.
18. Tang, D., W. Jin, D. Liu, J. Che, and Y. Yang, Siam deep feature KCF method and experimental study for pedestrian tracking. *Sensors*, 2023. **23**(1): p. 482.
19. Zhong, B., Y. Shen, Y. Chen, W. Xie, Z. Cui, H. Zhang, ... S. Peng, Online learning 3D context for robust visual tracking. *Neurocomputing*, 2015. **151**: pp. 710–718.

20. Zhang, W. and Z. Wang, FPFS: Filter-level pruning via distance weight measuring filter similarity. *Neurocomputing*, 2022. **512**: pp. 40–51.
21. Wang, Z., F. Li, G. Shi, X. Xie, and F. Wang, Network pruning using sparse learning and genetic algorithm. *Neurocomputing*, 2020. **404**: pp. 247–256.
22. Lian, Y., P. Peng, and W. Xu, Filter pruning via separation of sparsity search and model training. *Neurocomputing*, 2021. **462**: pp. 185–194.
23. Zhu, J. and J. Pei, Progressive kernel pruning with saliency mapping of input-output channels. *Neurocomputing*, 2022. **467**: pp. 360–378.
24. Chen, S.-B., Y.-J. Zheng, C.H. Ding, and B. Luo, SIECP: Neural network channel pruning based on sequential interval estimation. *Neurocomputing*, 2022. **481**: pp. 1–10.
25. Ma, X., G. Li, L. Liu, H. Liu, and X. Wang, Accelerating deep neural network filter pruning with mask-aware convolutional computations on modern CPUs. *Neurocomputing*, 2022. **505**: pp. 375–387.
26. Wang, L., W. Huang, M. Zhang, S. Pan, X. Chang, and S.W. Su, Pruning graph neural networks by evaluating edge properties. *Knowledge-Based Systems*, 2022. **256**: p. 109847.
27. Tian, G., J. Chen, X. Zeng, and Y. Liu, Pruning by training: A novel deep neural network compression framework for image processing. *IEEE Signal Processing Letters*, 2021. **28**: pp. 344–348.
28. Camci, E., M. Gupta, M. Wu, and J. Lin, Qlp: Deep q-learning for pruning deep neural networks. *IEEE Transactions on Circuits and Systems for Video Technology*, 2022. **32**(10): pp. 6488–6501.
29. De, T., P. Giri, A. Mevawala, R. Nemani, and A. Deo, Explainable AI: A hybrid approach to generate human-interpretable explanation for deep learning prediction. *Procedia Computer Science*, 2020. **168**: pp. 40–48.
30. Heuillet, A., F. Couthouis, and N. Díaz-Rodríguez, Explainability in deep reinforcement learning. *Knowledge-Based Systems*, 2021. **214**: p. 106685.
31. Kenny, E.M. and M.T. Keane, Explaining Deep Learning using examples: Optimal feature weighting methods for twin systems using post hoc, explanation-by-example in XAI. *Knowledge-Based Systems*, 2021. **233**: p. 107530.
32. Samek, W., G. Montavon, S. Lapuschkin, C.J. Anders, and K.-R. Müller, Explaining deep neural networks and beyond: A review of methods and applications. *Proceedings of the IEEE*, 2021. **109**(3): pp. 247–278.
33. Bai, X., X. Wang, X. Liu, Q. Liu, J. Song, N. Sebe, and B. Kim, Explainable deep learning for efficient and robust pattern recognition: A survey of recent developments. *Pattern Recognition*, 2021. **120**: p. 108102.
34. Terziyan, V. and O. Vitko, Explainable AI for Industry 4.0: Semantic representation of deep learning models. *Procedia Computer Science*, 2022. **200**: pp. 216–226.
35. He, L., N. Aouf, and B. Song, Explainable deep reinforcement learning for UAV autonomous path planning. *Aerospace Science and Technology*, 2021. **118**: p. 107052.
36. Jain, R., A. Kumar, A. Nayyar, K. Dewan, R. Garg, S. Raman, and S. Ganguly, Explaining sentiment analysis results on social media texts through visualization. *Multimedia Tools and Applications*, 2023: pp. 1–17.
37. Sun, Y., F. Haghighat, and B.C. Fung, Trade-off between accuracy and fairness of data-driven building and indoor environment models: A comparative study of pre-processing methods. *Energy*, 2022. **239**: p. 122273.

38. Petch, J., S. Di, and W. Nelson, Opening the black box: The promise and limitations of explainable machine learning in cardiology. *Canadian Journal of Cardiology*, 2022. **38**(2): pp. 204–213.
39. Brunetti, A., D. Buongiorno, G.F. Trotta, and V. Bevilacqua, Computer vision and deep learning techniques for pedestrian detection and tracking: A survey. *Neurocomputing*, 2018. **300**: pp. 17–33.
40. Bićanić, B., M. Oršić, I. Marković, S. Šegvić, and I. Petrović. Pedestrian tracking by probabilistic data association and correspondence embeddings. in *2019 22nd International Conference on Information Fusion (FUSION)*. 2019. IEEE.
41. Wong, P.K.-Y., H. Luo, M. Wang, P.H. Leung, and J.C. Cheng, Recognition of pedestrian trajectories and attributes with computer vision and deep learning techniques. *Advanced Engineering Informatics*, 2021. **49**: p. 101356.
42. Zheng, L., S. Zhao, Y. Zhang, and L. Yu, Thermal infrared pedestrian tracking using joint Siamese network and exemplar prediction model. *Pattern Recognition Letters*, 2020. **140**: pp. 66–72.
43. Gupta, N., S.K. Gupta, R.K. Pathak, V. Jain, P. Rashidi, and J.S. Suri, Human activity recognition in artificial intelligence framework: A narrative review. *Artificial Intelligence Review*, 2022. **55**(6): pp. 4755–4808.
44. Zhang, S., Y. Li, S. Zhang, F. Shahabi, S. Xia, Y. Deng, and N. Alshurafa, Deep learning in human activity recognition with wearable sensors: A review on advances. *Sensors*, 2022. **22**(4): p. 1476.
45. Pavel, M.I., S.Y. Tan, and A. Abdullah, Vision-based autonomous vehicle systems based on deep learning: A systematic literature review. *Applied Sciences*, 2022. **12**(14): p. 6831.
46. Rokonuzzaman, M., N. Mohajer, S. Nahavandi, and S. Mohamed, Review and performance evaluation of path tracking controllers of autonomous vehicles. *IET Intelligent Transport Systems*, 2021. **15**(5): pp. 646–670.
47. Ruslan, N.A.I., N.H. Amer, K. Hudha, Z.A. Kadir, S.A.F.M. Ishak, and S.M.F.S. Dardin, Modelling and control strategies in path tracking control for autonomous tracked vehicles: A review of state of the art and challenges. *Journal of Terramechanics*, 2023. **105**: pp. 67–79.

Chapter 12

Deep learning-based multi-object tracking

Advancement for intelligent video analysis

12.1 INTRODUCTION

Multi-object tracking (MOT) is one of the recent problems of computer vision which aims to track the trajectory of more than one object in a video sequence [1]. MOT is a two-step procedure in which initially the objects are detected using object detection algorithms. Next, the detected objects are tracked in the video sequences. MOT has a wide range of real-world applications that include crowd monitoring, autonomous car tracking, healthcare monitoring, and video surveillance.

MOT is similar in its applications to single-object tracking, but is more complex and tedious. Also, MOT requires the additional step of object detection to localize the objects in the initial frame [2]. This step determines the various categories of objects such as pedestrians, cars, and people, and tracks their trajectories in the entire video sequence by handling various tracking difficulties. Figure 12.1 illustrates the overview of the various steps involved in MOT. With advancements in technology, deep neural networks are explored in the field of MOT for developing efficient algorithms [3]. Recently, with the advancement in computing hardware processing, the power of algorithms has also increased. With the help of multi-core processors and graphics processing units (GPUs), the MOT can be achieved at a faster rate, making it suitable for real-time applications.

Recently, DL-based algorithms have been explored in MOT. These algorithms have provided efficient results by addressing the problems of occlusion, background clutters, pose variations, and many more. Deep neural networks extract high-level discriminative features rich in semantic and spatial information which are suitable to address these challenges. The MOT tracking algorithms are categorized into tracking-by-detection frameworks [4–7] and DL-based MOT frameworks [3, 8–10]. These tracking algorithms exploit various DL methodologies such as CNN either to extract spatial and temporal distinguishing features or to detect the target in the scene. In addition, object association and location estimation are determined using embedded learning [11].

DOI: 10.1201/9781003456322-12

Figure 12.1 Overview of steps in MOT.

The performance of MOT algorithms needs to be evaluated on various recent and publicly accepted evaluation metrics. For fair comparison and evaluation of state-of-the-art MOT, classical evaluation metrics such as MOTA (multi-object tracking accuracy), MOTP (multi-object tracking precision), IDS (number of ID switches), MT (mostly tracked) are a few essential which are utilized widely. Generally, these metrics are suitable to analyze the robustness of MOT algorithms in complex environmental variations.

Further, public benchmarks are proposed by various researchers to compare the performance of state-of-the-art MOT [12–14]. These data sets consist of videos containing complex and real-time situations suitable for the applicability of model to the real world. They are standardized for unified evaluation of the MOT algorithms. The detected objects are precisely annotated so that tracking errors can be minimized. The videos are split into training and test sequences so that performance can be tested on unseen videos.

In this chapter, we have summarized the latest advances in the MOT algorithms. The MOT algorithms are categorized into various classes so that salient features can be analyzed. The evaluation metrics are also elaborated for the calibration of state-of-the-art MOT. In addition, public MOT benchmarks are discussed so that a common and fair base for tracking algorithms evaluation can be provided. Finally, the details of the MOT algorithms are summarized in the last section.

12.2 MULTI-OBJECT TRACKING ALGORITHMS

MOT algorithms perform tracking in two stages. Initially, objects are identified in the first input frame. After that, the detected objects are tracked in the complete video sequences. Tracking can be performed by extracting features and analyzing the motion trajectory of all the detected objects. Also, the similarity scores and association between the objects belonging to the same class are computed for tracking. Based on this, MOT tracking algorithms are categorized into two classes: tracking-by-detection and DL-based MOT approaches. The salient features of various tracking algorithms are shown in Table 12.1. The exploited algorithms, features, and data sets are extracted so that their contribution to the MOT domain can be highlighted.

Table 12.1 Description of various frameworks under DL-MOT along with details of features extracted and data set exploited

SN	*References*	*Year*	*Methodology*	*Features*	*Data set*	*Summary*
1.	Chen et al. [4]	2021	Tracking by detection	Edge feature from spatio-temporal information and ORB feature	MOT15 and MOT17	Occlusion detection using object consistency and threshold classification methods.
2.	Yoon et al. [7]	2020	Tracking by detection	Deep features	MOT16 and MOT17	One-shot learning framework for tracking and detection of multiple objects.
3.	Alqaralleh et al. [15]	2020	DL and energy-efficient wireless multimedia sensor network	Recurrent neural network (RNN)	–	Tracking in two stages: fuzzy-based clustering and RNN-T based tracking algorithm.
4.	Xiang et al. [16]	2019	Deep learning	Deep features	MOT 2015 and MOT2016	RNN-based Bayesian filtering for predicting and estimating target state.
5.	Bae [8]	2020	Object model learning	RGB histogram, Shape, motion, and radar features	VS-PETS 2009	Object learning with data association methods and discriminative subspace learning for target tracking.
6.	Zhou et al. [10]	2019	Tracking by detection	Visual features	2DMOT15 and 2DMOT16	Confidence score-based weighting of tracklets for tracking objects using individual movement patterns and inter-object constraints.
7.	Lee and Kim [9]	2019	Pyramid Siamese network	Deep features and spatio-temporal motion	MOT17	Multi-level discriminative feature learning with spatio-temporal motion feature for MOT.
8.	Wan et al. [17]	2021	CNN	Deep features	MOT16 and MOT17	End-to-end deep learning for tracking objects for locating the objects.

9.	Ye et al. [18]	2022	Auxiliary trajectory association	Deep features	MOT16, MOT17, and MOT20	Lightweight Deep Appearance Embedding and a simulated occlusion strategy to address the occlusion challenge.
10.	Mahmoudi et al. [5]	2019	Tracking by detection (association)	Deep CNN features	MOT16	Grouping method based on affinity between the groups generated in previous frames.
11.	Singh and Srivastava [6]	2022	DL-MOT	Dense optical flow and heat map head	2DMOT15, MOT16, MOT17, and MOT20	Relative scale between the boundary boxes and relative position to calculate the relative distance between the objects for MOT.
12.	Lee et al. [19]	2021	Data association between frames	Edge and pose	MOT16 and MOT17	Tracker's performance improves by reducing data association time by minimizing the networks for feature aggregation and edge classification.
13.	Chu et al. [20]	2023	Spatial-temporal graph transformer	Visual features	MOT15, MOT16, MOT17, and MOT20	Encode the relationship between the visual features and the spatio-temporal relationship of the tracked targets.
14.	Shuai et al. [21]	2021	Siamese Network	Motion feature region-based and point based	MOT17	Model motion instance in the tracker's Siamese networks for multi-class tracking.
15.	Zhang et al. [22]	2022	Tracking by association	Re-ID features	MOT17 and MOT20	Track every bouncing box instead of selective ones. Similarity by tracklet to improve tracking by low score bounding boxes.
16.	Son et al. [23]	2017	Quadruplet CNN	Sequence-specific motion-aware position feature	2DMOT2015	Jointly learn object association and BB regression for an end-to-end trained unified network.

12.2.1 Tracking by detection

Primarily, tracking-by-detection MOT algorithms focuses on modeling the appearance variations of the detected objects to estimate the trajectory in video frames. Data association between the detected object is computed for tracking objects in the entire video frame. In this direction, the authors have proposed an edge multi-channel gradient for detecting the objects in the initial frame [4]. A special feature ORB key points are used for ensuring consistency between multiple objects. Matching algorithms along with the threshold classification method are modeled to obtain matching scores and address occlusion issues. Yoon et al. [7] have computed the similarity scores of the distance metrics to distinguish between the objects in videos. A neural network based on attentional learning is utilized to classify the unseen data samples using the labels identified in the video frames. The authors have presented an online association for MOT for maintaining low computations in a DL framework [5]. Affinity between the group objects is measured using a grouping method in the previous frame. Discriminative data-driven deep features such as primitive features from lower layers and higher layer features from the upper layers from a CNN are extracted.

In Ye et al. [18], the authors have proposed a lightweight deep appearance to model the association of trajectories for multiple objects. Affinity measure and discriminative appearance features are introduced to distinguish similar objects. Simulated occlusion strategy is also utilized to handle occlusion and to identify different targets in complex environmental conditions. However, Lee et al. [19] have utilized graph CNN to model the data association between the nodes. Similarity between the detection and tracking is computed in an end-to-end graph CNN network. The network parameters are minimized by designing three networks to achieve faster performance. Hungarian matching is used for updating the tracker's current state by combining the previous state to handle occlusion. On the other hand, the authors have presented the MOT tracker based on generic association between all the detected BB ignoring their detection scores [22].

Low score BB is used for determining the similarity between similar objects and removing background noise. During occlusion, high score BB does not match object tracklets which get impacted. Hence, low score BB is associated with the matched tracklets of the objects to recover them from occlusion. But Zhou et al. [10] have utilized inter-object relationships between the tracklets for generating discriminative features. High-confidence score tracklets correct the errors in low-confidence score tracklets to improve tracking accuracy. Trackers are trained end-to-end to adapt to tracking variations on target appearance and inter-object relations. Asymmetric pairwise term between the trajectories of the neighboring target is utilized to address the occlusion and missed detections. In summary, tracking-by-detection/association MOT frameworks are effective in addressing dense occlusion challenges. The affinity between the tracked objects is paramount for detecting similar objects of a class in the video stream.

12.2.2 Deep learning-based multi-object trackers (DL-MOT)

DL-MOT tracking aims to reduce the computational overheads of the tracking-by-detection/association framework by reducing the identity switching during tracking in complex crowded scenarios. In this direction, the authors have utilized the matching technique and relative position for calculating the relative scale and distance between the objects for MOT [6]. Matching technique is used for scaling up the indefinite number of objects in subsequent frames of a video while the relative position is for tracking the trajectories of those moving objects. Target-wise motion is extracted using flowNet-2 in an end-to-end learning framework. The authors have presented end-to-end learning with discriminative deep features [17]. The frames are provided as input and trajectories of the localized multiple objects are identified. Response maps are generated for the learned multiple objects and tracked in a video. Subnetworks are incorporated for motion displacement and handling occlusion.

In Xiang et al. [16], CNN is utilized for feature extraction, and LSTM network is designed for extracting motion information from the targets. Features from these two networks are integrated using a triple loss function in end-to-end deep metric learning. Target trajectories are reconstructed when either no objects are detected, or detected trajectories are inaccurate. The authors have proposed Quadruplet CNN for learning the associations between the detected objects [23]. Multi-task losses are used for learning the joint association between the object and BB regression. The network is trained in end-to-end metric learning for modeling the tracking association using minimax label propagation. The method can capture the data associations between the targets more accurately.

In Lee and Kim [9], the authors have proposed a pyramid Siamese network to enhance the structural simplicity for extracting the target's motion information. Spatio-temporal motion feature is also added to the tracker's architecture for discriminating similar objects so that trajectory estimation accuracy can be improved. Shuai et al. [21] have proposed a region-based Siamese network to minimize the structural loss in an end-to-end trainable network. The affinity between the detected objects is learned to improve the tracker's capability for estimating the target's movement. Occlusion is handled by considering the visibility confidence in future frames.

Graph transformers are proposed for tracking the trajectories of the objects [20]. Sparse weighted graphs are used for constructing the spatial graph transformer for encoding the layer. For long-term occlusion handling and low-score detection, cascade association frameworks are utilized. Computational processing is also catered for in efficient performance in comparison to conventional networks. In summary, DL-MOTs have shown accurate and effective tracking results in the presence of complex environmental conditions. Typical tracking variations such as severe occlusion and dense background clutter are addressed by incorporating additional mechanisms in the tracking strategy.

12.3 EVALUATION METRICS FOR PERFORMANCE ANALYSIS

Evaluation metrics are used for the estimation of performance of MOT in challenging scenarios. For a fair comparison of the trackers against state-of-the-art equivalents, clear metrics are utilized, as proposed by Bernardin and Stiefelhagen [24]. The clear metrics include MOTA (Multi-object tracking accuracy), MOTP (Multi-object tracking precision), ML (Mostly lost objects), MT (Mostly tracked objects), Identity switches (IDS), ID F1 score ($\mathrm{ID_{F1}}$), and FPS (frame per second). MOTA, ML, MT, and IDS are used for evaluating the computational performance whereas $\mathrm{ID_{F1}}$ and FPS are parameters to evaluate the processing speed. MOTA can be computed using Eq. (12.1).

$$\mathrm{MOTA} = 1 - \frac{F^{\mathbb{N}} + F^{\wp} + \mathrm{TID}^{\mathrm{S}}}{C_t^{\mathrm{G}}} \tag{12.1}$$

where $F^{\mathbb{N}}$ and $F^{\wp}$ are the total number of false positives and negatives in the entire video. TID^S is the total number of ID switches in the whole video and C_t^{G} is the total number of ground truth BB. MOTA can vary from $-\infty$ to 1 for an algorithm as the number of errors can be more than the number of ground truth BB. Hence, MOTA is expressed in percentages to eliminate confusion. Next, MOTP can be calculated using Eq. (12.2).

$$\mathrm{MOTP} = \frac{\sum_{f,i} \mathrm{BB}_{f,i}}{\sum_{f} m_f} \tag{12.2}$$

where, $\mathrm{BB}_{f,i}$ is the BB overlap between hypothesis i with its assigned ground truth BB and m_f is the number of matches in frame f.

$\mathrm{ID_{F1}}$ is the harmonic mean of identification precision and identification recall. It is computed using Eq. (12.3).

$$\mathrm{ID_{F1}} = \frac{2\,\mathrm{ID_{TP}}}{2\mathrm{ID_{TP}} + \mathrm{ID_{FP}} + \mathrm{ID_{FN}}} \tag{12.3}$$

where $\mathrm{ID_{TP}}$ denotes the sum of weights of accurate detection in the entire video. $\mathrm{ID_{FP}}$ and $\mathrm{ID_{FN}}$ are the sums of weights from the selected false positive IDs and false negative IDs, respectively.

Apart from the above performance metrics, ML is computed in percentage and represents the number of ground truth trajectories that are tracked correctly in less than 20% of the total frames. MT denotes the number of ground truth trajectories that are accurately tracked in at least 80% of the frames and it is also generally represented in percentage. IDS represents the number of times an object is tracked accurately, but sometimes the connected ID of the object gets changed. In addition, FPS is used to represent the processing speed of the MOT in terms of its dependency on hardware.

Primarily, the above-discussed metrics are used for computing the tracking performance for state-of-the-art MOT. These metrics are useful to provide a common base to compare the different MOT algorithms in terms of their power to address the tracking variations.

12.4 BENCHMARK FOR PERFORMANCE EVALUATION

Before the practical implication of the MOT trackers, there is a requirement for video MOT data sets that can test the tracker's performance in real-time scenarios. A lot of public tracking MOT benchmarks have been published in the last few years. In this section, the most recent and common MOT data sets are described. Table 12.2 shows the salient features and general description of commonly used MOT data sets. The data set attributes are also extracted to evaluate them in terms of efficiency to compare various MOT tracking algorithms.

Earlier, PETS2009 MOT data set is proposed by Ferryman and Shahrokni [25]. This data set contains videos captured by multiple cameras having more than one target. This data set provides crowded and complex sequences with a high density of people, suitable for real-world tracking estimation. PETS2009 is divided into multiple sub-data sets, specific to an objective and difficulty. The video sequences are having a resolution of 768 × 576 and 720 × 576 at 7 FPS.

- Leal-Taixé et al. [13] have proposed the MOT15 benchmark for object tracking, pedestrian tracking, single short-term tracking, and 3D reconstruction. The data set consists of standardized short-duration video sequences with potential challenges. The video sequences are captured with static as well as moving cameras. Videos have varied resolutions of 640 × 480, 768 × 576, 1920 × 1080, 1242 × 370, 1242 × 375, and 1920 × 1080 captured at varying FPS from 10 to 25. The data set is a collection of videos taken from previous data sets and some newly captured also. However, most of the video sequences are easy and not annotated carefully. Due to this, the performance of the MOT tracking algorithms cannot be evaluated properly.
- To address the limitation of the MOT15 data set, Milan et al. [26] have proposed the MOT 16/17 data set which is a collection of some existing videos and newly captured videos. All the videos are annotated carefully to avoid inconsistency between the video sequences. The data set consists of a wide range of videos with different viewpoints, varying lighting conditions, and high target density. The videos

Table 12.2 Description of various MOT benchmarks

SN	*References*	*Year*	*Data set*	*No. of videos*	*Data set attributes*	*Train/ Test Split*	*Summary*
1.	Ferryman and Ellis [25]	2009	PETS2009	8	Varying resolution and 7 FPS	Yes	Combination of three data sets categorized to address crowd monitoring in tedious scene complexity.
2.	Leal-Taixé et al. [13]	2015	MOT15	22	Multiple classes, varying resolutions, and FPS	Yes	Combination of public video sequences along with new sequences for tracker's performance comparison.
3.	Milan et al. [26]	2016	MOT16/17	14	11k+ images, 11 classes, varying resolutions, and FPS	Yes	Data set with a variety of classes and more précised annotations.
4.	Dendorfer et al. [12]	2020	MOT20	8 videos from 3 different scene	13k+ images, Varying resolution, 8 classes, 25 FPS	Yes	Data sets with varying tracking challenges are suitable for testing the tracker's performance in crowded scenes.
5.	Sun et al. [14]	2022	Dancetrack	100	105k+ images, Diversified scene, 20 FPS	Yes	Multiple dance sequences with varying attributed challenges.
6.	Pedersen et al. [27]	2023	BrackishMOT	98	671 frames, 8 object classes, 15 FPS	Yes	Underwater tracking data sets for tracking aquatic life in complex real-time situations.

The data sets are arranged in ascending order of year of publication.

are captured with stationary as well as moving cameras. The videos have a resolution of 1920 × 1080 and 640 × 480 at varying FPS of 30, 25, and 14 suitable for real-time scenarios.

- Recently, the MOT20 data set was proposed, consisting of 8 newly captured videos with complex tracking challenges [12]. Average duration of the videos is 2 mins, with 1920 × 1080, 1173 × 880, 1654 × 1080, 1920 × 734, and 1545 × 1080, captured at 25 FPS. Data sets consists of indoor, outdoor, day, and night videos. Video sequences have non-motorized vehicles, static persons, and crowds of persons.
- Sun et al. [14] have proposed MOT DanceTrack large data set having video sequences with diverse non-linear dancing motion patterns. The data set consists of 100 video sequences with an average length of 52 FPS captured at 20 FPS. Video sequences are captured indoors and outdoors with various dancing forms such as street, pop, classical, and group, along with sports scenarios. Videos are annotated with the help of a commercial annotation tool to avoid inconsistency. Frames with full occlusion are not annotated. The data set provides videos to capture frequent pose variations and complex motion patterns.
- Pedersen et al. [27] have proposed the underwater data set BrackishMOT for tracking fish. The data set consists of 98 wild video sequences captured in underwater scenarios and categorized into 6 different classes. Videos have small to medium-sized fish along with various objects to track underwater. The data set is a combination of real and synthetic underwater sequences. Synthetic underwater sequences are generated to evaluate the performance of underwater MOT tracking algorithms.

To summarize, public MOT data sets play a crucial role in evaluating the performance of MOT tracking algorithms. They provide a common base so that the tracking capability of the tracking algorithms can be simulated and compared by addressing real-time tracking limitations. These data sets are rich in complex tracking problems suitable for crowd monitoring, pedestrian tracking, pose tracking, and many more applications of MOT.

12.5 APPLICATION OF MOT ALGORITHMS

MOT has provided a lot of applications in various fields of computer vision. MOT can provide solutions to high-level computer vision problems. Primarily, MOT aims to track multiple objects in complex surroundings. The main application of MOT is in object segmentation, identification, 3D human pose estimation, pedestrian tracking, surveillance, and motion prediction.

- The authors have pre-trained the fully connected CNN for image classification by extracting the important properties of CNN features [28]. Features from different layers are observed to have different discriminative powers and are used for discriminating the target from the background.

- A collaborative algorithm is proposed for extracting the multiple-segment tracks from the object [29]. Initially, object detectors were used for generating multiple BBs for tracking objects. Then, each BB is transformed into a pixel-wise segment using alternate shrinking and expansion segmentation. Object disappearance and reappearance are addressed by refining the segment tracks.
- MOT is used for object tracking and segmentation using dense pixel-level annotations [30]. Objects are segmented in the first frame of the video sequences using semi-automatic annotations.
- Person re-identification from multiple cameras using semi-supervised attribute deep learning in MOT is proposed by Su et al.[31]. The authors have proposed a three-stage framework to identify the same person from multiple cameras. Deep CNN network is first trained on an independent data set, then fine-tuned on another data set, and finally, the attribute labels are combined to obtain the final results.
- Multiple people are tracked, and the pose is estimated in unconstrained videos [32]. Body joint locations of the targets are formulated using spatio-temporal grouping. The body part relations are sparsified using a body part relationship graph to speed up the process of pose estimation.
- 3D animation of human pose and action is estimated using a supervised multi-person framework known as AnimePose [33]. Initially, objects are detected and segmented in the video, and depth maps are estimated. Videos are converted from 2D to 3D for target localization, trajectory prediction, and human pose tracking.
- VoxelTrack is proposed for MOT 3D pose estimation and tracking from multiple cameras [34]. Multiple views are projected to capture the object identification, and then a feature vector is computed for each voxel by averaging the heatmaps obtained from body joints. The approach is robust even when the target is fully occluded.
- Extreme motion between two interacting persons is predicted by exploiting a cross interaction mechanism [35]. Future motion is predicted by observing the historical information and dependencies between the two persons by formulating collaborative tasks.
- PI-Net is proposed for multi-person monocular 3D pose estimation using deep networks [36]. Interaction between the various pose estimates is the analysis by understanding the dependency between the current and the future pose into a recurrent architecture. Pose estimation provides useful information for people interaction, physical therapy, and many more real-time applications.
- Automated video surveillance for MOT is explored for crowd counting and detecting suspicious activity [37]. Complex environmental situations for crowd monitoring are addressed using inverse transform

and median filtering. Targets are detected by the head torso and tracking is performed using the Kalman filter adopting Jaccard similarity and normalized cross-correlation.

- Intelligent video surveillance along with pedestrian detection is performed using the iterative particle repropagation method [38]. A greedy approach is employed for obtaining the matching score for data association and model update.

To summarize, MOT has provided solutions to many real-time problems. With the advancement in technology, improvement in tracking accuracy and precision, reliability, and MOT methodologies have gained momentum in the past few years. But the lack of data set availability, the authenticity of synthetic data, and evaluation on a single data set restricts the practical implications of these methods.

12.6 LIMITATIONS OF EXISTING MOT ALGORITHMS

A lot of work has been proposed under MOT for providing a robust solution [40]. However, certain limitations have restricted the practical implications of MOT in real-time applications such as surveillance, crowd monitoring, and autonomous vehicle tracking.

- Primarily, the detection quality of the algorithms is of paramount importance. The number of false negative detections should be minimized for reliable tracking results. High detection quality must be ensured by improving the performance parameters of the tracking algorithms.
- Discriminative feature extraction is necessary to discriminate similar objects in MOT. Strong appearance models with reliable motion information must be integrated to keep track of the target trajectories. Deep features rich in semantic and spatial information of the target must be extracted using deep neural networks such as CNN to enhance the quality of tracking algorithms.
- Trackers based on the tracking-by-detection framework are based on the detection hypothesis for tracking objects. However, in these trackers' the description of detection algorithms is hidden. Basic specifications of the detection are not disclosed, and so the practical implication of these trackers is inhibited.
- Emphasis should be given to reducing the complexity of the MOT algorithms. Algorithms that are complicated and complex may provide high accuracy, but are not feasible solutions for real-time tracking problems. Training a large number of parameters in the DL-based MOT algorithms reduces the processing speed of the tracker. The tradeoff between algorithm complexity and accuracy should be chosen

carefully. Accuracy should not be sacrificed for reducing the complexity of the algorithms.
- MOT algorithms are specific to particular scenarios. An algorithm may perform better on one data set but may not generate satisfying results on the other data sets. Training of algorithms on data sets having either a small number of videos or easy sequences causes the overfitting problem. This problem may lead to the failure of the MOT algorithm in real-time scenarios.

To summarize, all these prevent the practical applications of MOT in the real world. Steps must be taken to address these limitations and provide robust and reliable tracking results.

12.7 SUMMARY

MOT tracking algorithms detect targets in the first frame and provide them with IDs so that identified objects can be tracked in the entire video sequence. However, there is a requirement for a robust detector that can mitigate the number of false detections to improve tracking accuracy. Complex MOT applications such as pedestrian tracking and crowd monitoring demand higher precision in terms of quality of the tracking algorithms. Hence, online trained detectors with better efficiency need to be exploited their applicability in MOT.

DL-MOT tracking is keenly investigated for performance improvements by integrating high-end feature extraction networks. The increase in complexity of the network improves accuracy, but the processing speed of the tracker is reduced. The computation complexity of the DL-MOT is paramount providing fast-tracking results. For this, the attentional mechanism and transformers are explored in the MOT.

Most of the MOT algorithms fail to address the varying lighting conditions, contrast, illumination variations, and dense occlusions. In addition, the algorithms are specific to one kind of problem and fail to perform when scene attributes change. Recently, modern MOT trackers have been exploited to address the problem but still there is a requirement for adaptive MOT trackers that can understand the scene contexts and can generate reliable tracking results.

Public benchmark MOT data sets are published to provide a common base for MOT trackers evaluation. However, the quality of videos in the data sets is not good, and the annotations are also incorrect. Due to the limited availability of data in some MOT applications such as pose estimation, either the synthetic data is generated from the existing data, or the authors have used self-generated data. Self-generated data is not reliable enough to ensure unbiased tracking performance [39].

MOT has provided solutions to many recent computer vision tasks that include object segmentation, identification, surveillance, and 3D object tracking. It can also be explored for tracking wild animals in the forest, underwater fish tracking, traffic scene monitoring, and many more applications. MOT algorithms should be flexible and generic enough to track different kinds of objects in different tracking situations.

REFERENCES

1. Ciaparrone, G., F.L. Sánchez, S. Tabik, L. Troiano, R. Tagliaferri, and F. Herrera, Deep learning in video multi-object tracking: A survey. *Neurocomputing*, 2020. **381**: pp. 61–88.
2. Xu, Y., X. Zhou, S. Chen, and F. Li, Deep learning for multiple object tracking: A survey. *IET Computer Vision*, 2019. **13**(4): pp. 355–368.
3. Pal, S.K., A. Pramanik, J. Maiti, and P. Mitra, Deep learning in multi-object detection and tracking: State of the art. *Applied Intelligence*, 2021. **51**: pp. 6400–6429.
4. Chen, J., Z. Xi, C. Wei, J. Lu, Y. Niu, and Z. Li, Multiple object tracking using edge multi-channel gradient model with ORB feature. *IEEE Access*, 2021. **9**: pp. 2294–2309.
5. Mahmoudi, N., S.M. Ahadi, and M. Rahmati, Multi-target tracking using CNN-based features: CNNMTT. *Multimedia Tools and Applications*, 2019. **78**: pp. 7077–7096.
6. Singh, D. and R. Srivastava, An end-to-end trained hybrid CNN model for multi-object tracking. *Multimedia Tools and Applications*, 2022. **81**(29): pp. 42209–42221.
7. Yoon, K., J. Gwak, Y.-M. Song, Y.-C. Yoon, and M.-G. Jeon, Oneshotda: Online multi-object tracker with one-shot-learning-based data association. *IEEE Access*, 2020. **8**: pp. 38060–38072.
8. Bae, S.-H., Online multi-object tracking with visual and radar features. *IEEE Access*, 2020. **8**: pp. 90324–90339.
9. Lee, S. and E. Kim, Multiple objects tracking via feature pyramid Siamese networks. *IEEE Access*, 2019. **7**: pp. 8181–8194.
10. Zhou, H., W. Ouyang, J. Cheng, X. Wang, and H. Li, Deep continuous conditional random fields with asymmetric inter-object constraints for online multi-object tracking. *IEEE Transactions on Circuits and Systems for Video Technology*, 2019. **29**(4): pp. 1011–1022.
11. Wang, G., M. Song, and J.-N. Hwang, Recent advances in embedding methods for multi-object tracking: A survey. arXiv preprint arXiv:2205.10766, 2022.
12. Dendorfer, P., H. Rezatofighi, A. Milan, J. Shi, D. Cremers, I. Reid, ... L. Leal-Taixé, Mot20: A benchmark for multi-object tracking in crowded scenes. arXiv preprint arXiv:2003.09003, 2020.
13. Leal-Taixé, L., A. Milan, I. Reid, S. Roth, and K. Schindler, Motchallenge 2015: Towards a benchmark for multi-target tracking. arXiv preprint arXiv:1504.01942, 2015.
14. Sun, P., J. Cao, Y. Jiang, Z. Yuan, S. Bai, K. Kitani, and P. Luo. Dancetrack: Multi-object tracking in uniform appearance and diverse motion. in *Proceedings of the IEEE/CVF Conference on Computer Vision and Pattern Recognition*. 2022.

15. Alqaralleh, B.A., S.N. Mohanty, D. Gupta, A. Khanna, K. Shankar, and T. Vaiyapuri, Reliable multi-object tracking model using deep learning and energy efficient wireless multimedia sensor networks. *IEEE Access*, 2020. **8**: pp. 213426–213436.
16. Xiang, J., G. Zhang, and J. Hou, Online multi-object tracking based on feature representation and Bayesian filtering within a deep learning architecture. *IEEE Access*, 2019. 7: pp. 27923–27935.
17. Wan, X., J. Cao, S. Zhou, J. Wang, and N. Zheng, Tracking beyond detection: learning a global response map for end-to-end multi-object tracking. *IEEE Transactions on Image Processing*, 2021. **30**: pp. 8222–8235.
18. Ye, L., W. Li, L. Zheng, and Y. Zeng, Lightweight and deep appearance embedding for multiple object tracking. *IET Computer Vision*, 2022. **16**(6): pp. 489–503.
19. Lee, J., M. Jeong, and B.C. Ko, Graph convolution neural network-based data association for online multi-object tracking. *IEEE Access*, 2021. **9**: pp. 114535–114546.
20. Chu, P., J. Wang, Q. You, H. Ling, and Z. Liu. Transmot: Spatial-temporal graph transformer for multiple object tracking. in *Proceedings of the IEEE/CVF Winter Conference on Applications of Computer Vision*. 2023.
21. Shuai, B., A. Berneshawi, X. Li, D. Modolo, and J. Tighe. Siammot: Siamese multi-object tracking. in *Proceedings of the IEEE/CVF Conference on Computer Vision and Pattern Recognition*. 2021.
22. Zhang, Y., P. Sun, Y. Jiang, D. Yu, F. Weng, Z. Yuan, ... X. Wang. Bytetrack: Multi-object tracking by associating every detection box. in *Computer Vision–ECCV 2022: 17th European Conference, Tel Aviv, Israel, October 23–27, 2022, Proceedings, Part XXII*. 2022. Springer.
23. Son, J., M. Baek, M. Cho, and B. Han. Multi-object tracking with quadruplet convolutional neural networks. in *Proceedings of the IEEE Conference on Computer Vision and Pattern Recognition*. 2017.
24. Bernardin, K. and R. Stiefelhagen, Evaluating multiple object tracking performance: The clear mot metrics. *EURASIP Journal on Image and Video Processing*, 2008. **2008**: pp. 1–10.
25. Ferryman, J. and A. Shahrokni. Pets2009: Data set and challenge. in *2009 Twelfth IEEE International Workshop on Performance Evaluation of Tracking and Surveillance*. 2009. IEEE.
26. Milan, A., L. Leal-Taixé, I. Reid, S. Roth, and K. Schindler, MOT16: A benchmark for multi-object tracking. arXiv preprint arXiv:1603.00831, 2016.
27. Pedersen, M., D. Lehotský, I. Nikolov, and T.B. Moeslund, BrackishMOT: The Brackish multi-object tracking data set. arXiv preprint arXiv:2302.10645, 2023.
28. Wang, L., W. Ouyang, X. Wang, and H. Lu. Visual tracking with fully convolutional networks. in *Proceedings of the IEEE International Conference on Computer Vision*. 2015.
29. Koh, Y.J. and C.-S. Kim. CDTS: Collaborative detection, tracking, and segmentation for online multiple object segmentation in videos. in *2017 IEEE International Conference on Computer Vision (ICCV)*. 2017. IEEE.
30. Voigtlaender, P., M. Krause, A. Osep, J. Luiten, B.B.G. Sekar, A. Geiger, and B. Leibe. Mots: Multi-object tracking and segmentation. in *Proceedings of the IEEE/CVF Conference on Computer Vision and Pattern Recognition*. 2019.

31. Su, C., S. Zhang, J. Xing, W. Gao, and Q. Tian. Deep attributes-driven multi-camera person re-identification. in *Computer Vision–ECCV 2016: 14th European Conference, Amsterdam, The Netherlands, October 11-14, 2016, Proceedings, Part II 14*. 2016. Springer.
32. Insafutdinov, E., M. Andriluka, L. Pishchulin, S. Tang, E. Levinkov, B. Andres, and B. Schiele. Arttrack: Articulated multi-person tracking in the wild. in *Proceedings of the IEEE Conference on Computer Vision and Pattern Recognition*. 2017.
33. Kumarapu, L. and P. Mukherjee, Animepose: Multi-person 3d pose estimation and animation. *Pattern Recognition Letters*, 2021. **147**: pp. 16–24.
34. Zhang, Y., C. Wang, X. Wang, W. Liu, and W. Zeng, Voxeltrack: Multi-person 3d human pose estimation and tracking in the wild. *IEEE Transactions on Pattern Analysis and Machine Intelligence*, 2022. **45**(2): pp. 2613–2626.
35. Guo, W., X. Bie, X. Alameda-Pineda, and F. Moreno-Noguer. Multi-person extreme motion prediction. in *Proceedings of the IEEE/CVF Conference on Computer Vision and Pattern Recognition*. 2022.
36. Guo, W., E. Corona, F. Moreno-Noguer, and X. Alameda-Pineda. Pi-net: Pose interacting network for multi-person monocular 3d pose estimation. in *Proceedings of the IEEE/CVF Winter Conference on Applications of Computer Vision*. 2021.
37. Shehzed, A., A. Jalal, and K. Kim. Multi-person tracking in smart surveillance system for crowd counting and normal/abnormal events detection. in *2019 International Conference on Applied and Engineering Mathematics (ICAEM)*. 2019. IEEE.
38. Choi, J.W., D. Moon, and J.H. Yoo, Robust multi-person tracking for real-time intelligent video surveillance. *Etri Journal*, 2015. **37**(3): pp. 551–561.
39. Kumar, A., Walia, G.S., and Sharma, K. (2020). Recent trends in multicue based visual tracking: A review. *Expert Systems with Applications*, **162**: p. 113711.
40. Kumar, A., R. Jain, V. A. Devi, & A. Nayyar, (Eds.). *Object Tracking Technology: Trends, Challenges, Impact, and Applications*, 2023. Springer.

Index

Pages in **bold** refer tables.